INTERNAL COMBUSTION ENGINES

Edited by **Kazimierz Lejda** and **Paweł Woś**

Internal Combustion Engines
http://dx.doi.org/10.5772/2806
Edited by Kazimierz Lejda and Paweł Woś

Contributors

Wladyslaw Mitianiec, Yoshihito Yagyu, Hideo Nagata, Nobuya Hayashi, Hiroharu Kawasaki, Tamiko Ohshima, Yoshiaki Suda, Seiji Baba, Eliseu Monteiro, Marc Bellenoue, Julien Sottton, Abel Rouboa, Simón Fygueroa, Carlos Villamar, Olga Fygueroa, Artur Jaworski, Hubert Kuszewski, Kazimierz Lejda, Adam Ustrzycki, Teresa Donateo, Mudassar Abbas Rizvi, Qarab Raza, Aamer Iqbal Bhatti, Sajjad Zaidi, Mansoor Khan, Sorin Rațiu, Corneliu Birtok-Băneasă, Yuki Kudoh

Published by InTech

Janeza Trdine 9, 51000 Rijeka, Croatia

Publishing Process Manager Marina Jozipovic
Technical Editor InTech DTP team
Cover InTech Design team

Additional hard copies can be obtained from orders@intechopen.com

Internal Combustion Engines, Edited by Kazimierz Lejda and Paweł Woś
p. cm.
ISBN 978-953-51-0856-6

For EU product safety concerns:
IN TECH d.o.o., Prolaz Marije Krucifikse Kozulić 3, 51000 Rijeka, Croatia,
info@intechopen.com or visit our website at intechopen.com.

Meet the editors

Professor Kazimierz Lejda graduated in mechanical engineering in 1971 from Cracow University of Technology where he also earned his PhD degree in internal combustion engine engineering in 1983. In 1993 he obtained a postdoctoral qualification from Kiev Institute of Vehicles and Roads (Ukraine) within automotive vehicle and combustion engine engineering. He has stayed employed in Rzeszów University of Technology since 1973; he got the chair of Department of Automotive Vehicles and Internal Combustion Engines in 1994, and was honoured with professorship in 2011. His scientific interests cover development of injection systems for greening engines and improving their performances. Prof. Lejda is a member of Polish Academy of Sciences and other domestic and international scientific organizations, and is a very prolific industrial and scientific researcher and publisher; he has drawn up more than 365 items including 7 books. He also reveals great contribution to young scientists' and engineers' education.

Paweł Woś is an assistant professor in Rzeszów University of Technology, where he graduated in mechanical engineering in 1995, and earned his PhD degree in combustion engine engineering in 2003. Within 2005 to 2006 he served as a post-doc fellow in Wayne State University in Detroit (USA) where he worked on advanced combustion processes in ICE. He continues the studies with application of variable compression ratio (VCR) technology. He is the author of more than 45 papers and scientific reports.

Contents

Preface

Internal combustion engines (ICE) are the main sources of powering for almost all road vehicles, yet many other machines too. Being under strength development for a number of years, they have already reached a relatively high level of technical excellence and now they also produce acceptable output parameters. Still, they are not devoid of drawbacks. Harmful exhaust emissions can be pointed as the most important here. This problem is the main focus of interest for automotive researchers and engineers. Continuous decrease of exhaust emission limits additionally intensifies their efforts to produce more green engines and vehicles. On the other hand, rapid development of road transportation and the growth of end-users' demands toward more and more comfortable, durable, reliable and fuel-saving vehicles unceasingly calls for improvements in engine design and technology.

Despite many attempts, replacing the internal combustion engine with other, but equally effective power source still fails. Therefore, extensive works on the improvement of internal combustion engines should be carried out and the results need to be widely published.

As the answer to above expectations, this book on internal combustion engines brings out few chapters on the research activities through the wide range of current engine issues. The first section groups combustion-related papers including all research areas from fuel delivery to exhaust emission phenomena. The second one deals with various problems on engine design, modeling, manufacturing, control and testing. Such structure should improve legibility of the book and helps to integrate all singular chapters as a logical whole.

We wish to thank InTech Publisher and are especially pleased to express same thanks to Ms. Viktorija Žgela for giving us an invitation and opportunity to be editors of the book on internal combustion engines. Distinctive thanks are also due to Ms. Romana Vukelić and Ms. Marina Jozipović, and Publishing Process Staff for their help in coordinating the reviews, editing and printing of the book.

Kazimierz Lejda and **Paweł Woś**
Rzeszów University of Technology,
Poland

Section 1

Engine Fuelling, Combustion and Emission

Chapter 1

Factors Determing Ignition and Efficient Combustion in Modern Engines Operating on Gaseous Fuels

Wladyslaw Mitianiec

Additional information is available at the end of the chapter

http://dx.doi.org/10.5772/48306

1. Introduction

Recently in automotive industry the applying of gaseous fuels and particularly compressed natural gas both in SI and CI engines is more often met. However application of CNG in the spark ignition internal combustion engines is more real than never before. There are known many designs of the diesel engines fuelled by the natural gas, where the gas is injected into inlet pipes. Because of the bigger octane number of the natural gas the compression ratio of SI engines can be increased, which takes effect on the increase of the total combustion efficiency. In diesel engines the compression ratio has to be decreased as a result of homogeneity of the mixture flown into the cylinder. Such mixture cannot initiate the self-ignition in traditional diesel engines because of higher value of CNG octane number. Direct injection of the compressed natural gas requires also high energy supplied by the ignition systems. A natural tendency in the development of the piston engines is increasing of the air pressure in the inlet systems by applying of high level of the turbo-charging or mechanical charging. Naturally aspirated SI engine filled by the natural gas has lower value of thermodynamic efficiency than diesel engine. The experiments conducted on SI engine fuelled by CNG with lean homogeneous mixtures show that the better solution is the concept of the stratified charge with CNG injection during the compression stroke. The presented information in the chapter is based on the own research and scientific work partly described in scientific papers. There is a wider discussion of main factors influencing on ignition of natural gas in combustion engines, because of its high temperature of ignition, particularly at high pressure. The chapter presents both theoretical considerations of CNG ignition and experimental work carried out at different air-fuel ratios and initial pressure.

Gas engines play more and more important role in automotive sector. This is caused by decreasing of crude oil deposits and ecologic requirements given by international institutions concerning to decreasing of toxic components in exhaust gases. Internal combustion engines should reach high power with low specific fuel consumption and indicate very low exhaust gas emission of such chemical components as hydrocarbons, nitrogen oxides, carbon monoxide and particularly for diesel engines soot and particulate matters. Chemical components which are formed during combustion process depend on chemical structure of the used fuel. Particularly for spark ignition engines a high octane number of fuel is needed for using higher compression ratio which increases the thermal engine efficiency and also total efficiency.

2. Thermal and dynamic properties of gas fuels

The mixture of the fuel and oxygen ignites only above the defined temperature. This temperature is called as the ignition temperature (self-ignition point). It is depended on many internal and external conditions and therefore it is not constant value. Besides that for many gases and vapours there are distinguished two points: lower and higher ignition points (detonation boundary). These two points determine the boundary values where the ignition of the mixture can follow. The Table 1 presents ignition temperatures of the stoichiometric mixtures of the different fuels with the air.

Fuel	Ignition temperature [°C]	Fuel	Ignition temperature [°C]
Gasoline	350 - 520	Brown coal	200 - 240
Benzene	520 - 600	Hard coal atomised	150 - 220
Furnace oil	≈ 340	Coking coal	≈ 250
Propane	≈ 500	Soot	500 - 600
Charcoal	300 - 425	Natural gas	≈ 650
Butane (n)	430	City gas	≈ 450
Furnace oil EL	230 - 245	Coke	550 - 600

Table 1. Ignition temperatures of the fuels in the air (mean values)

The combustion mixture, which contains the fuel gas and the air, can ignite in strictly defined limits of contents of the fuel in the air. The natural gas consists many hydrocarbons, however it includes mostly above 75% of methane. For the experimental test one used two types of the natural gas:

1. the certified model gas G20 which contains 100% of methane compressed in the bottles with pressure 200 bar at lower heat value 47.2 – 49.2 MJ/m^3
2. the certified model gas G25 that contain 86% of methane and 14% of N2 at lower heat value 38.2 – 40.6 MJ/m^3.

The natural gas delivered for the industry and households contains the following chemical compounds with adequate mean mass fraction ratios: methane - 0.85, ethane - 0.07, propane - 0.04, n-butane - 0.025, isobutene - 0.005, n-pentane - 0.005, isopentane - 0.005.

Because the natural gas contains many hydrocarbons with changeable concentration of the individual species the heat value of the fuel is not constant. It influences also on the ignition process depending on lower ignition temperature of the fuel and energy induced by secondary circuit of the ignition coil. For comparison in Table 2 the ignition limits and temperatures for some technical gases and vapours in the air at pressure 1.013 bars are presented. The data show a much bigger ignition temperature for the natural gas (640 – 670 °C) than for gasoline vapours (220°C). For this reason the gasoline-air mixture requires much lower energy for ignition than CNG-air mixture. However, higher pressure during compression process in the engine with higher compression ratio in the charged SI engine causes also higher temperature that can induce the sparking of the mixture by using also a high-energy ignition system. Because of lower contents of the carbon in the fuel, the engines fuelled by the natural gas from ecological point of view emit much lower amount of CO_2 and decreases the heat effect on our earth.

Till now there are conducted only some laboratory experiments with the high-energy ignition system for spark ignition engines with direct CNG injection. There are known the ignition systems for low compressed diesel engines fuelled by CNG by the injection to the inlet pipes.

Type of gas	Chemical formula	Normalized density (air = 1)	Ignition limits in the air (% volumetric)	Ignition temperature in the air [°C]
Gasoline	~C_8H_{17}	0.61	0.6 - 8	220
Butane (n)	C_4H_{10}	2.05	1.8 – 8.5	460
Natural gas H		0.67	5 - 14	640
Natural gas L		0.67	6 - 14	670
Ethane	C_2H_6	1,047	3 – 12.5	510
Ethylene	C_2H_4	1,00	2.7 - 34	425
Gas propane-butane 50%		1.79	2 - 9	470
Methane	CH_4	0.55	5 - 15	595
Propane	C_3H_8	1.56	2,1 – 9.5	470
City gas I		0.47	5 - 38	550
City gas II		0.51	6 - 32	550
Carbon monoxide	CO	0.97	12.5 - 74	605
Hydrogen	H_2	0.07	4 - 76	585
Diesel oil		0.67	0.6 – 6.5	230

Table 2. Ignition limits and ignition temperatures of the most important technical gases and vapours in the air at pressure 1,013 bar

Composition and properties of natural gas used in experimental tests are presented in Table 3.

No	Parameter	Nomenclature or symbol	Unit	Value
1	Combustion heat	Q_c	[MJ/Nm³] [MJ/kg]	39,231 51,892
2	Calorific value	W_d	[MJ/Nm³] [MJ/kg]	35,372 46,788
3	Density in normal conditions	ρ_g	[kg/Nm³]	0,756
4	Relative density	Δ	-	0,586
5	Coefficient of compressibility	Z	-	0,9980
6	Wobbe number	W_B	[MJ/Nm³]	51,248
7	Stoichiometric constant	L_o	[Nm^3_{fuel}/Nm^3_{air}]	9,401
8	CO_2 from the combustion	-	[Nm^3/Nm^3]	0,999

Table 3. Properties of the natural gas used in experimental research

3. Fuelling methods and ignition in gas diesel engines

Several fuelling methods of the natural gas are applied in modern compression ignition engines, where the most popular are the following cases:

- delivering the gas fuel into the inlet pipes by mixing fuel and air in the special mixer
- small pressure injection of gaseous fuel into the pipe and ignition of the mixture in the cylinder by electric spark
- high pressure direct injection of gaseous fuel particularly in high load engine

There are given the reasons of decreasing of compression ratio in two first methods and the aim of application of gaseous fuels in CI engines (lowering of CO_2, elimination of soot and better formation of fuel mixture). Applying of the two first methods decreases the total engine efficiency in comparison to standard diesel engine as a result of lowering of compression ratio and needs an additional high energetic ignition system to spark disadvantages of application of gas fuel in CI engines. Figure 1 presents an example of variation of heat release of dual fuel naturally aspirated 1-cylinder compression ignition engine Andoria 1HC102 filled by CNG and small amount of diesel oil as ignition dose. This type of engine is very promising because of keeping the same compression ratio and obtaining of higher total efficiency. NG in gaseous forms is pressured into the inlet pipe, next flows by the inlet valve into the cylinder. During compression stroke small dose of diesel oil is delivered by the injector into the combustion chamber as an ignition dose. Because ignition temperature of diesel oil is lower than that of natural gas the ignition start begins from the outer sides of diesel oil streams. In a result of high temperature natural gas the combustion process of the natural gas begins some degrees of CA later. The cylinder contains almost homogenous mixture before the combustion process and for this reason burning of natural gas mixture proceeds longer than that of diesel oil. Figure 1 presents simulation results carried out for this engine in KIVA3V program.

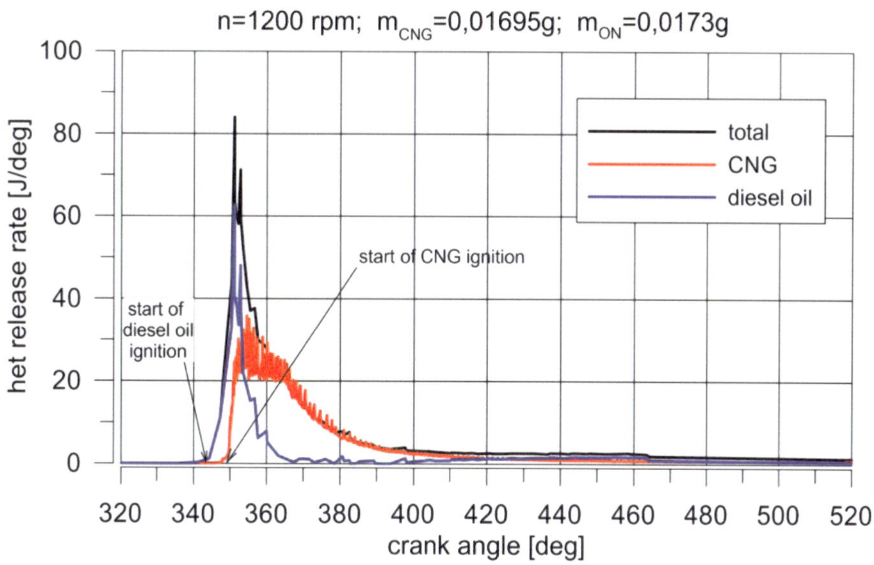

Figure 1. Heat release rate in dual fuel Andoria 1HC102 diesel engine fuelled by CNG and ignition dose of diesel oil (index ON- diesel oil, CNG – natural gas)

At higher load of diesel engine with dual fuel a higher mass of natural gas is delivered into the cylinder with the same mass of ignition diesel oil. In order to obtain the same air excess coefficient λ as in the standard diesel engine the following formula was used:

$$\lambda_{eq} = \frac{m_{air}}{m_{\text{do}}\left(\frac{A}{F}\right)_{\text{do}} + m_{\text{CNG}}\left(\frac{A}{F}\right)_{\text{CNG}}} \tag{1}$$

where: m_{air} - mass of air in the cylinder,
m_{do} - mass of diesel oil dose,
m_{CNG} - mass of CNG in the cylinder,
A/F - stoichiometric air-fuel ratio.

At assumed the filling coefficient $\eta_v = 0{,}98$ and charging pressure at the moment of closing of the inlet valve p_o = 0,1 MPa and charge temperature T_o = 350 K, the air mass delivered to the cylinder with piston displacement V_s amounts:

$$m_{air} = V_s \frac{\varepsilon}{\varepsilon - 1} \frac{p_o}{RT_o} \eta_v - m_{\text{CNG}} \tag{2}$$

At the considered dual fuelling the calculated equivalent air excess coefficients after inserting into eq. (2) and next into eq. (1) amounted, respectively: 1) at n = 1200 rpm - λ_z = 2,041, 2) at n=1800 rpm - λ_z = 1,359, 3) at n=2200 rpm - λ_z = 1,073.

Variation of the mass of natural gas in the dual fuel Andoria 1HC102 diesel engine at rotational speed 2200 rpm is shown in Figure 2. The principal period of combustion process of the natural gas lasted about 80 deg CA and its ignition began at TDC. In the real engine the diesel oil injection started at 38 deg CA BTDC.

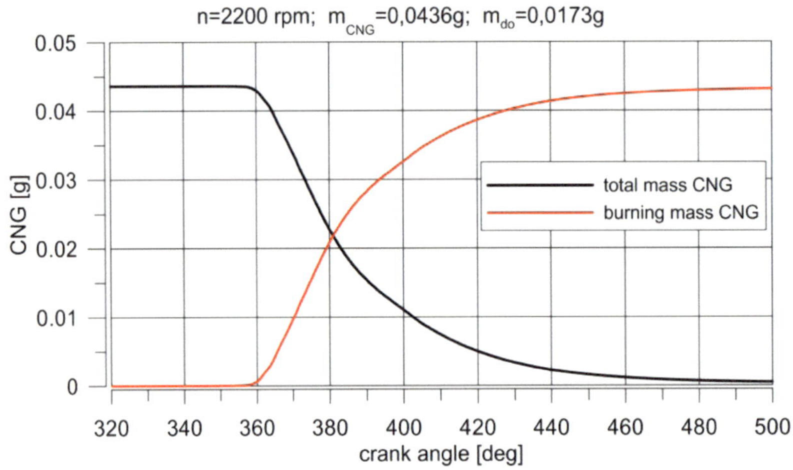

Figure 2. Mass variation of natural gas in Andoria 1HC102 diesel engine fuelled by CNG and ignition dose of diesel oil (index do- diesel oil, CNG – natural gas)

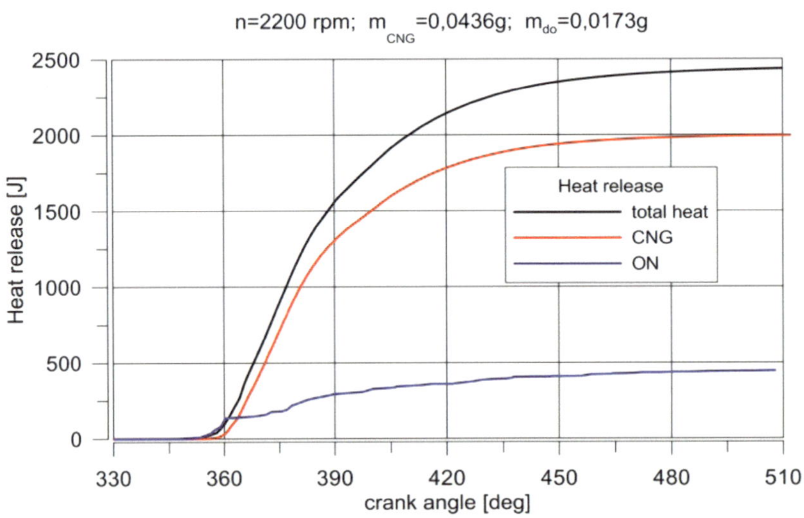

Figure 3. Heat release in dual fuel Andoria 1HC102 diesel engine fuelled by CNG and ignition dose of diesel oil (index do- diesel oil, CNG – natural gas)

Heat release from the both fuels (CNG and diesel oil) is shown in Figure 3 for the same engine at rotational speed 2200 rpm. Total heat released during combustion process results mainly on higher burning mass of the natural gas. The ignition process in the gas diesel engines with the ignition dose of diesel oil differs from other systems applied in modified engines fuelled by natural gas delivered into the inlet pipe and next ignited by the spark plug. The initiation of combustion process in CNG diesel engines with spark ignition is almost the same as in the spark ignition engines.

4. Ignition conditions of natural gas mixtures

The flammability of the natural gas is much lower than gasoline vapours or diesel oil in the same temperature. At higher pressure the spark-over is more difficulty than at lower pressure. During the compression stroke the charge near the spark plug can be determined by certain internal energy and turbulence energy. Additional energy given by the spark plug at short time about 2 ms increases the total energy of the mixture near the spark plug. The flammability of the mixture depends on the concentration of the gaseous fuel and turbulence of the charge near the spark plug. Maximum of pressure and velocity of combustion process in the cylinder for given rotational speed depend on the ignition angle advance before TDC (Figure 4).

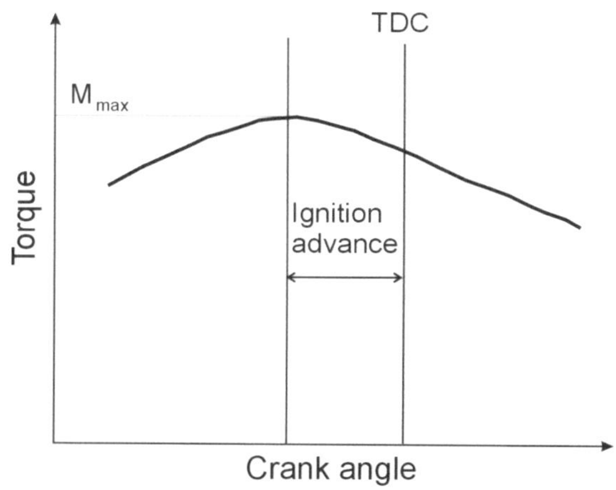

Figure 4. Influence of ignition angle advance on the engine torque

The beginning of the mixture combustion follows after several crank angle rotation. While this period certain chemical reactions follow in the mixture to form the radicals, which can induce the combustion process. The energy in the spark provided a local rise in temperature of several thousand degrees Kelvin, which cause any fuel vapour present to be raised above its auto-ignition temperature. The auto-ignition temperature determines the possibility of the break of the hydrocarbon chains and the charge has sufficient internal energy to oxidize the carbon into CO_2 and water in the vapour state. Immediately, after the beginning of combustion (ignition point) the initial flame front close to the spark plug moves in a radial direction into the space of the combustion chamber and heats the unburned layers of air-fuel mixture surrounding it.

For the direct injection of CNG for small loads of the engine in stratified charge mode the burning of the mixture depends on the pressure value at the end of compression stroke and on the relative air-fuel ratio. These dependencies of the CNG burning for different mixture composition and compression ratio are presented in Figure 5 [15]. The burning of CNG mixture can occur in very small range of the compression pressure and lean mixture composition and maximum combustion pressure reaches near 200 bars. For very lean mixtures and higher compression ratios the misfire occurs, on the other hand for rich mixtures and high compression ratios the detonation is observed. During the cold start-up the ignition process of the CNG mixture is much easier than with gasoline mixture because of whole fuel is in the gaseous state. Today in the new ignition systems with electronic or capacitor discharge the secondary voltage can reach value 40 kV in some microseconds.

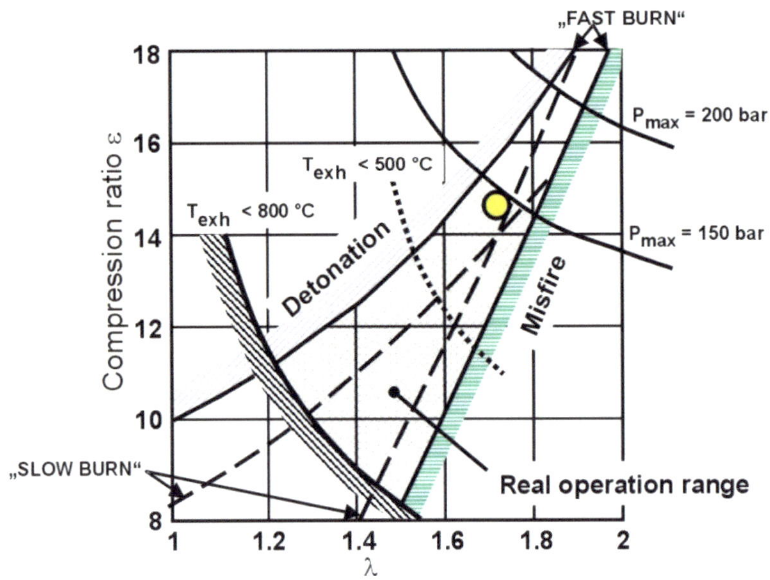

Figure 5. The range of combustion limits for lean CNG mixture [3]

The higher voltage in the secondary circuit of the transformer and the faster spark rise enable that the sparking has occurred even when the spark plug is covered by liquid gasoline. With fuelling of the engine by CNG the sparking process should occur in every condition of the engine loads and speeds. However, at higher compression ratio and higher engine charging the final charge pressure increases dramatically in the moment of ignition and this phenomenon influences on the sparking process.

5. Electric and thermal parameters of ignition

On the observation and test done before on the conventional ignition systems, the higher pressure of the charge in the cylinder requires also higher sparking energy or less the gap of the electrodes in the spark plug. The chemical delay of the mixture burning is a function of the pressure, temperature and properties of the mixture and was performed by Spadaccini [12] in the form:

$$\Delta\tau_z = 2.43\cdot 10^{-9} p^{-2}\cdot \exp[41560/(R\cdot T)] \tag{3}$$

where: p - pressure [bar], T - temperature [K] and R - gas constant [(bar cm^3)/(mol K)].

The simplest definition of this delay was given by Arrhenius on the basis of a semi-empirical dependence:

$$\Delta\tau_z = 0.44 \cdot 10^{-3} \cdot p^{-1.19} \exp\left[\frac{4650}{T}\right] \tag{4}$$

where p is the charge pressure at the end of the compression process [daN/cm²].

Experimental and theoretical studies divide the spark ignition into three phases: breakdown, arc and glow discharge. They all have particular electrical properties. The plasma of temperature above 6000 K and diameter equal the diameter of the electrodes causes a shock pressure wave during several microseconds. At an early stage a cylindrical channel of ionization about 40 μm in diameter develops, together with a pressure jump and a rapid temperature rise. Maly and Vogel [10] showed that an increase in breakdown energy does not manifest itself a higher kernel temperatures, instead the channel diameter causing a larger activated-gas volume. Since the ratio between the initial temperature of the mixture and the temperature of the spark channel is much smaller than unity, the diameter d of the cylindrical channel is given approximately by the following expression:

$$d = 2\left[(\gamma - 1)\frac{E_{bd}}{\gamma \cdot \pi \cdot h \cdot p}\right]^{1/2} \tag{5}$$

where γ is ratio of the specific heats, h is the spark plug gap and p pressure. E_{bd} represents the breakdown energy to produce the plasma kernel. Ballal and Lefebvre [6] considered the following expression for the breakdown voltage U_{bd} and total spark energy E_t:

$$U_{bd} = \frac{2{,}8 \cdot 10^5 \cdot p \cdot h}{5{,}5 - \ln(p \cdot d)} \qquad E_t = \int_0^{t_i} V \cdot I \cdot dt \tag{6}$$

One assumed, that the charge is isentropic conductive and the field attains a quasi-steady state (no time influence). Knowing the potential of the electromagnetic field Φ and electrical conductivity σ the following equation can be used [12]:

$$\text{div}(-\sigma \cdot \text{grad}\Phi) = 0 \tag{7}$$

After a forming of the plasma between the electrodes the heat source $\dot{q}_e$ in the mixture can be calculated directly from the electrical current in the secondary coil circuit I, which changes during with time:

$$\dot{q}_e = \sigma \frac{I^2}{\left(\int_0^R 2\pi r \sigma(r,z) dr\right)^2} \tag{8}$$

where r and z are the coordinates of the ionization volume.

At leaner homogenous mixture the discharging of the energy by spark plug leads sometimes to the misfire and increasing of the hydrocarbons emission. At stratified charge for the same total air-fuel ratio the sparking of the mixture can be improved by turning the injected fuel directly near spark plug at strictly defined crank angle rotation depending on the engine speed. The energy involved from the spark plug is delivered to the small volume near spark plug. The total energy, which is induced by the spark plug is a function of the voltage and current values in the secondary circuit of the ignition coil and time of the discharge. On the other hand, values of voltage *U* and current *I* change in the discharge time and total energy induced by the coil can be expressed as a integral of voltage *U*, current *I* and time *t*:

$$E_{ign} = \int_0^{\tau} U \cdot I \cdot dt \tag{9}$$

where τ is the time of current discharge by the secondary circuit of the ignition coil. Integration of the measurement values of voltage and current in the secondary circuit of the coil gives the total electric energy to the mixture charge near spark plug. The total internal energy of the mixture near the spark plug increases in the period t = 0..τ and according to the energy balance in the small volume the temperature of the charge in this region continuously increases.

The modern conventional ignition system can give the burning energy e_{burn} = 60 mJ at the secondary voltage 30 kV and burning current i_{burn} = 70 mA during 1.8 ms. In practice a required value of the secondary voltage of the ignition system is calculated from the following formula:

$$U_2 = 4700 \cdot (a \cdot \varepsilon)^{0.718} \tag{10}$$

where: U_2 - secondary voltage [V],
a - gap between electrodes of the spark plug,
ε - compression ratio.

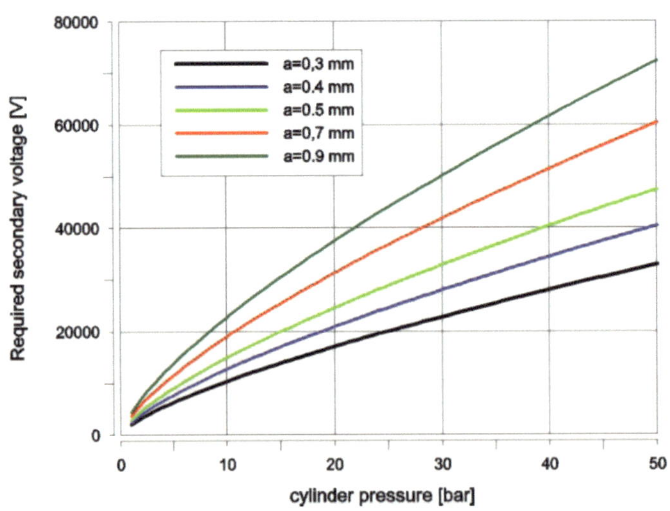

Figure 6. The secondary voltage as a function of compression pressure and electrode's gap

For lower gaps and compression ratios the secondary voltage can be decreased. The required secondary voltage as a function of compression pressure is presented in Figure 6 for different gaps of spark plug electrodes from 0.3 to 0.9 mm.

If one assumes that the electrical energy E is delivered during period τ to a certain small volume V near spark plug with the temperature of the charge T_1 and pressure p_1 and concentration of CNG fuel adequate to the air excess coefficient λ, it is possible to calculate the change of the charge temperature in this space. On the basis of the law of gas state and balance of energy the specific internal energy u of the charge in the next step of calculation is defined.

$$u_i = u_{i-1} + dE \tag{11}$$

where i is the step of calculations and dE is the energy delivered from the spark plug in step time $d\tau$. The internal energy is function of the charge mass m and temperature T, where mass m in volume V is calculated from the following dependency:

$$m = \frac{p_1 \cdot V}{R \cdot T_1} \tag{12}$$

and gas constant R is calculated on the mass concentration g of the n species in the mixture. Mass of the charge consists of the fuel mass m_f and air mass m_a, which means:

$$m = m_a + m_f \tag{13}$$

For the mixture that contains only air and fuel (in our case CNG), the equivalent gas constant is calculated as follows:

$$R = \sum_1^n g_i \cdot R_i = g_a \cdot R_a + g_f \cdot R_f \tag{14}$$

In simple calculations the local relative air-fuel ratio λ is obtained from the local concentration of air and fuel:

$$\lambda = \frac{m_a}{K \cdot m_f} \tag{15}$$

where K is stoichiometric coefficient for a given fuel. For the CNG applied during the experiments K=16.04 [kg air/kg CNG]. At assumption of the relative air-fuel ratio λ the masses of fuel m_f and air m_a can be obtained from the following formulas:

$$m_f = \frac{m}{\lambda K + 1} \qquad m_a = m \frac{\lambda K}{\lambda K + 1} \tag{16}$$

After substitution of the fuel and air masses to the equation (10) the equivalence gas constant R is defined only if the λ is known.

$$R = \frac{1}{\lambda K + 1}\left(\lambda \cdot K \cdot R_a + R_f\right) \tag{17}$$

For whole volume V the internal energy at the beginning of the ignition is defined as:

$$U_1 = m \cdot c_v \cdot T_1 = \frac{p_1 \cdot V}{R \cdot T_1} \cdot c_v \cdot T_1 = \frac{p_1 \cdot V}{R} \cdot c_v \tag{18}$$

The charge pressure during compression process increases as function of the crank angle rotation from p_1 to p. When one knows the engine's stroke S and diameter D of the cylinder and compression ratio ε it is possible to determine the change of pressure from start point to another point. If the heat transfer will be neglected the pressure change in the cylinder can be obtained from a simple formula as a function of time t and engine speed n (rev/min):

$$\frac{dp}{dt} = -\frac{30}{\pi}\frac{k-1}{n \cdot V_c}\left(\frac{k}{k-1}\frac{dV_c}{dt}\right) \tag{19}$$

where V_c is volume of the cylinder at crank angle φ and k is specific heat ratio (c_p/c_v).

For simplicity of calculations it was assumed that during compression stroke the specific heat ratio for small period is constant ($k \approx 1.36$) and cylinder volume changes with kinematics of crank mechanism. Delivery of electrical energy to the local volume results on the increase of local internal energy and changing of temperature T, which can be determined from the following energy equation:

$$m \cdot c_v \cdot T_i = m \cdot c_v \cdot T_{i-1} + de \quad \text{or} \quad m \cdot c_v \cdot \frac{dT}{dt} = \frac{de}{dt} \tag{20}$$

The electrical energy can be performed in a different way: with constant value during time τ (rectangular form or according to the reality in a triangular form as shown in Figure 7.

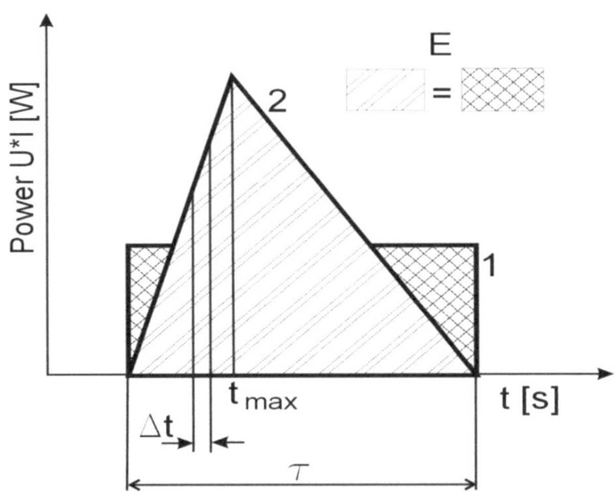

Figure 7. Variation of electrical power from spark plug

If the total electrical energy amounts E and duration of sparking lasts τ (1.8 ms) then for the first case the local power is E/τ for whole period τ of the sparking. For the second case electrical power from the spark plug changes and for the first period can be expressed as:

$$N_I = \frac{t}{t_{max}} \frac{2E}{\tau} \tag{21}$$

For the second period the electrical power can be determined as follows:

$$N_{II} = \frac{1 - t/\tau}{1 - t_{max}/\tau} \cdot \frac{2 \cdot E}{\tau} \tag{22}$$

The temperature of the charge near the spark plug during the period τ is computed as follows:

$$dT = \frac{1}{m \cdot c_v} N(t) \cdot dt \tag{23}$$

For the first case (rectangular form) of variation of electrical power the change of the charge temperature is computed from the following dependency:

$$dT = \frac{1}{m \cdot c_v} \cdot \frac{E}{\tau} \cdot dt \tag{24}$$

For the second case (triangular form of power) the temperature of the local charge is calculated as follows:

a. 1st period

$$dT = \frac{1}{m \cdot c_v} \cdot \frac{t}{t_{max}} \cdot \frac{2E}{\tau} \cdot dt \tag{25}$$

b. 2nd period

$$dT = \frac{1}{m \cdot c_v} \cdot \frac{1 - \frac{t}{\tau}}{1 - \frac{t_{max}}{\tau}} \cdot \frac{2 \cdot E}{\tau} \cdot dt \tag{26}$$

At assuming of specific volumetric heat c_v as constant in a small period τ the temperature of the local charge is simply obtained by integration of given above equations as function of time t ($t = 0 .. \tau$)

1. $$T = T_1 + \frac{E}{m \cdot c_v} \frac{t}{\tau} \tag{27}$$

$$2a. \qquad T = T_1 + \frac{E}{m \cdot c_v} \cdot \frac{\tau}{t_{max}} \cdot \left(\frac{t}{\tau}\right)^2 \tag{28}$$

$$2b. \qquad T = C + \frac{1}{m \cdot c_v} \cdot \frac{2 \cdot E}{1 - \frac{t_{max}}{\tau}} \cdot \frac{t}{\tau}\left(1 - \frac{t}{2\tau}\right) \tag{29}$$

The constant C is calculated for the initial conditions for $t/\tau = t_{max}/\tau$ with the end temperature for 1st period as an initial temperature for 2nd period. The three cases are performed in a non-dimensional time t/τ. Because compression stroke in 4-stroke engine begins usually φ_a=45° CA ABDC and thus the cylinder volume [3] can be calculated at φ_i crank angle as follows:

$$V_a = \frac{V_s}{\varepsilon - 1} + \frac{V_s}{2}\left(1 + \frac{\delta}{4} - \cos(180 - \varphi_i) - \frac{\delta}{4}\cos 2(180 - \varphi_i)\right) \tag{30}$$

The simple calculations of the increment of the local temperature in the region of the spark plug were done at certain assumptions given below: swept volume of the cylinder - 450 cm^3, compression ratio – 12, crank constant δ - 0.25, diameter of sparking region - 1 mm, height of sparking region - 1 mm, closing of inlet valve - 45° CA ABDC, start angle of ignition - 20° CA BTDC.

For calculation the air-gas mixture was treated as an ideal gas (methane CH_4 and air at λ=1.4). Two ignition systems were considered with ignition energy 40 and 60 mJ at assumption of:

1. constant sparking power (rectangular form) in period τ=2 ms
2. variable sparking power (triangular form) in period τ=2 ms.

The results of calculations are performed in Figure 8 for those two ignition systems, respectively. It was assumed that compression process begins after closing of the inlet valve with constant coefficient of compression politrope k=1.36.

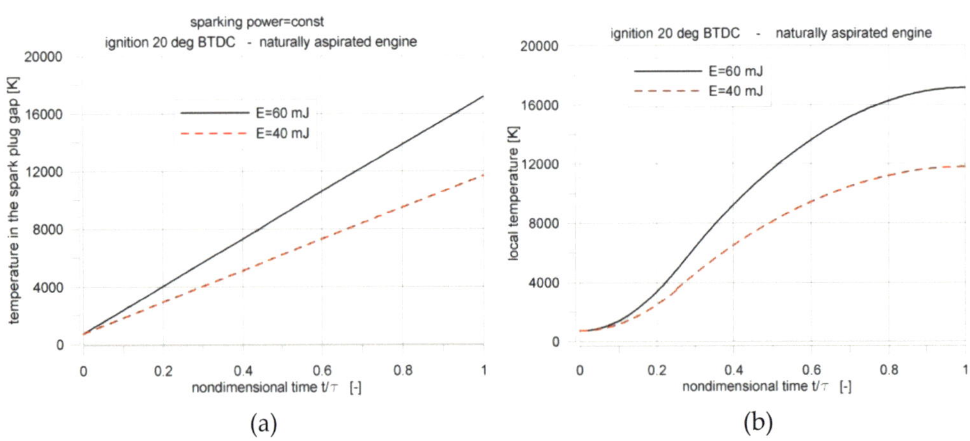

Figure 8. Increment of the local temperature in the region of the spark plug for two ignition systems: a) with constant sparking power, b) with variable sparking power (triangular form)

In the moment of the sparking start the pressure in the cylinder amounts 1.577 MPa at temperature 726 K. Theoretical consumption of the air for combustion of 1 Nm^3 of the natural gas amounts 9.401 Nm^3. For given concentration of the air and fuel (CNG) in the mixture the gas constant is R=296.9 J/(kg K) and calculated mass of charge in the region amounts 0.465e-8 kg. As shown in both figures the final temperature in the region is the same for two considered variations of power. If the volume of the sparking region decreases the local temperature will increase, however ignition of the mixture depends on concentration of the fuel in the air. The final temperature does not depend on the shape of the ignition power during sparking but only on the total energy released during the sparking. In the gap of the electrodes at ignition energy 60 mJ a mean temperature amounts almost 17000 K after 2 ms and at 40 mJ amounts 12000 K. This is enough to ignite the mixture.

6. Determination of thermal efficiency

Only a small part of the delivered energy from the second circuit is consumed by gaseous medium, which is observed by increase of the temperature ΔT and thus also internal energy E_i. The thermal efficiency of the ignition system is defined as ratio of the increase of internal energy and energy in the secondary circuit of the ignition coil:

$$\eta_{th} = \frac{\Delta E_i}{E_2} = \frac{\Delta E_i \cdot E_1}{E_2 \cdot E_1} = \eta_o \cdot \eta_e \tag{31}$$

where E_1 is the energy in the primary circuit and η_0 is the total efficiency and η_e is the electric efficiency of the ignition system. The increase of the internal energy in volume V with initial pressure p_1 can be determined as follows:

$$\Delta E_i = m \cdot c_v \cdot \Delta T \tag{32}$$

Assuming a constant mass and individual gas constant R, the temperature after ignition can be defined from the gas state equation. At small change of the gas temperature from T_1 to T_2 the volumetric specific heat c_v has the same value. In such way it is possible to determine the increase of the internal energy:

$$\Delta E_i = \frac{p_1 \cdot V}{R \cdot T_1} \cdot c_v \cdot (T_2 - T_1) = \frac{p_1 \cdot V}{R \cdot T_1} \cdot c_v \cdot \left(T_1 \cdot \frac{p_2}{p_1} - T_1 \right) \tag{33}$$

After simplification this equation takes the form:

$$\Delta E_i = \frac{V}{R} \cdot c_v \cdot (p_2 - p_1) = \frac{V}{R} \cdot c_v \cdot \Delta p \tag{34}$$

The increase of the internal energy depends on the sparking volume, gas properties and a pressure increment in this volume. Because of constant volume and known R and c_v the unknown value is only the increment of the pressure Δp. The direct method of measurement is using the pressure piezoelectric transducer with big sensitivity and with high limit of static pressure. For that case we have used the sensor PCB Piezotronic 106B51 (USA) with the following parameters:

- Measurement range (for ±5V output) 35 kPa
- Maximum pressure (step) 690 kPa
- Maximum pressure (static) 3448 kPa
- Sensitivity (±15%) 145 mV/kPa

For that sensor the amplifier Energocontrol VibAmp PA-3000 was used. The filling of the chamber with fixing of the spark plug and transducer is presented in Figure 9. The additional (medium) chamber with capacity 200 cm^3 is filled under given pressure (shown on the manometer) from the pressure bottle. The caloric chamber is filled from this medium chamber by the special needle valves. After sparking the chamber was emptied by opening the other needle valve. The needle valves were used in order to decrease the dead volume in the pipes connecting the chamber. The total volume was measured by filling the chamber by water and amounts 4,1 cm^3.

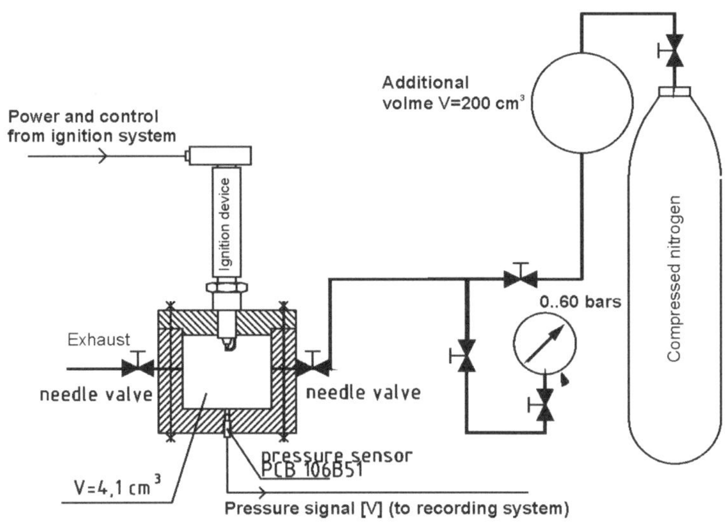

Figure 9. Scheme of the direct measurement pressure in the caloric chamber

The target of the tests was to determine the amount of thermal energy delivered do the charge in the chamber after the sparking; it means the measurements of the pressure increment in function of initial pressure. For one point of each characteristic we carried out 10 measurements. For the tests two types of electrodes were used: the normal with 2.8 mm width and the "thin" with 25% cross-section of the first type. The measurements were carried out in nitrogen and air at initial pressure in the chamber corresponded to ambient conditions (over pressure 0 bar) and at 25 bars. For the "thin" electrodes there is observed a bigger increment of the pressure than while using the spark plug with normal electrodes both at low as at high initial pressure, despite the delivered energy from the secondary circuit of the coil is almost the same. Increment of pressure inside the chamber caused by energy delivered from spark plug is shown in Figure 10 for initial pressure 1 bar and 25 bars and by application of the spark plug with "thin" and "thick" electrodes.

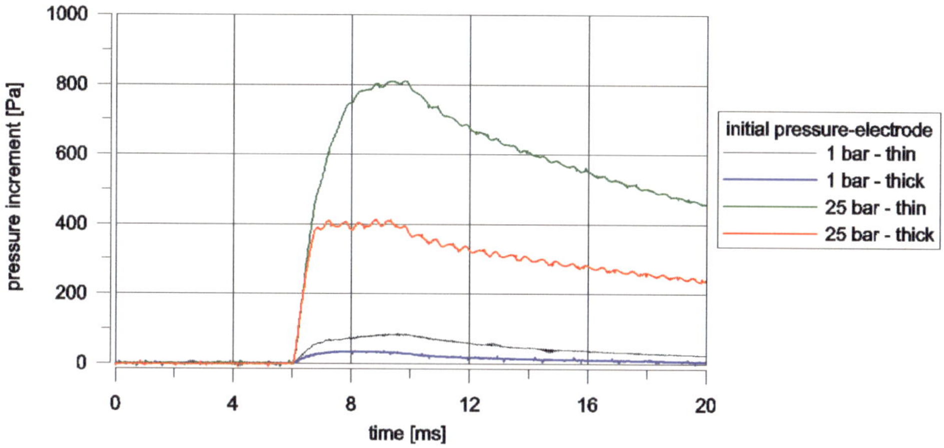

Figure 10. Pressure increment in caloric chamber filled by nitrogen at initial pressure 1 and 25 bars by application of spark plug with "thin" and "thick" electrodes

The duration of the sparking lasted about 4 ms and after this time the decrement of the pressure is observed which is caused by heat exchange with walls of the caloric chamber. In every case at the end of ignition process the sudden increase of secondary voltage takes place. The current in the secondary circuit of the ignition coil increases rapidly to about 80 mA after signal of the ignition and then decreases slowly during 4 ms to zero as one shows in Figure 11 for all considered cases.

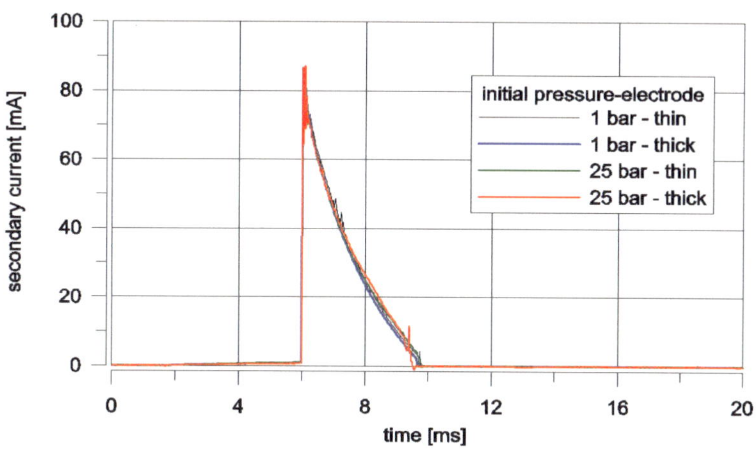

Figure 11. Secondary current in the coil during the ignition in the caloric chamber filled by nitrogen at initial pressure 1 and 25 bar by application of spark plug with "thin" and "thick" electrodes

Variation of voltage in the secondary circuit is shown in Figure 12. For the considered ignition coil one reaches maximum voltage 3000 V in the case of higher initial pressure 30 bar. In every case at the end of ignition process the sudden increase of secondary voltage takes place. Thermal energy delivered to the spark plug (in the secondary circuit) was determined by integration of instant electric power (multiplication of current and voltage) with small time step. For the case with "thin" electrodes and at 1 bar the thermal energy amounts only 0,89 mJ and thus the thermal efficiency is about η_{th} = 1,29% (Figure 13). For normal electrodes at the same pressure the thermal energy is very lower 0,36 mJ which causes a small thermal efficiency η_{th} = 0,51%.

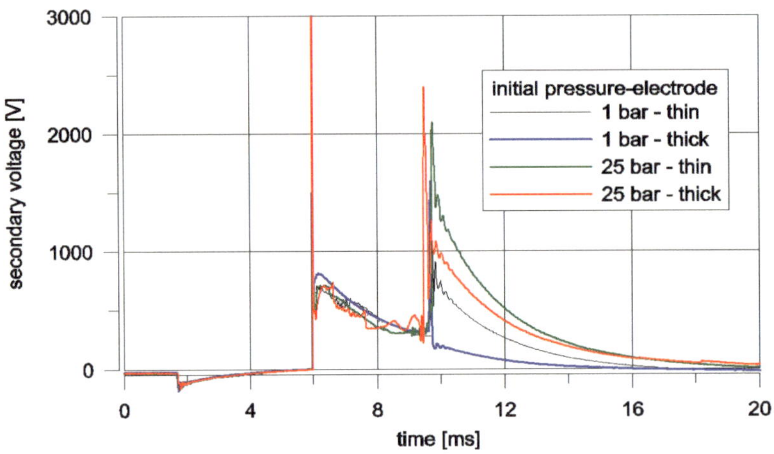

Figure 12. Secondary voltage in the coil during the ignition in the caloric chamber filled by nitrogen at initial pressure 1 and 25 bars by application of spark plug with "thin" and "thick" electrodes

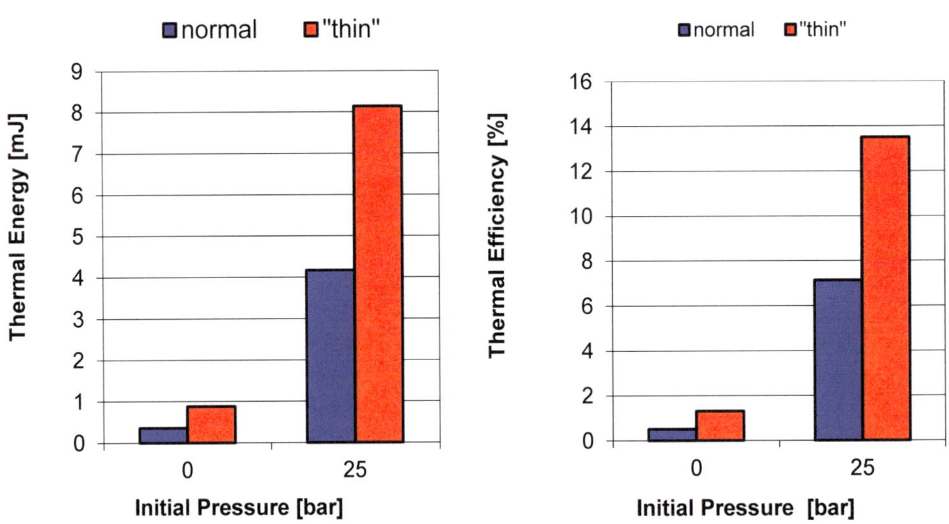

Figure 13. The comparison of the thermal energy and thermal efficiency for spark plug with normal and "thin" electrodes at two initial positive gauge pressures

The thermal energy and thermal efficiency increases with the increase of the initial pressure. For the case with "thin" electrodes of the spark plug the thermal efficiency amounts 13.49%, on the other hand for normal electrodes only 6.93%. The tests were done for five ignition systems from BERU at different initial pressure (0 – 25 bars) and linear approximation variations of the thermal efficiencies are shown in Figure 14. With increasing of the pressure in the caloric chamber much more energy is delivered from the electric arc to the gas. The measurements of the pressure increase during spark ignition were carried out also for the air and the same pressures. Figure 16 presents the increase of secondary voltage in the ignition coil with increasing of initial pressure in the caloric chamber. For nitrogen and leaner mixtures a higher secondary voltage in the coil was measured.

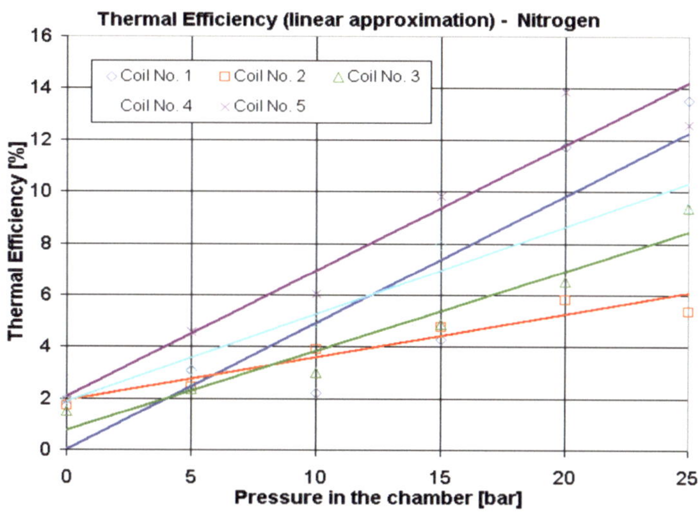

Figure 14. Thermal efficiency of five tested ignition systems

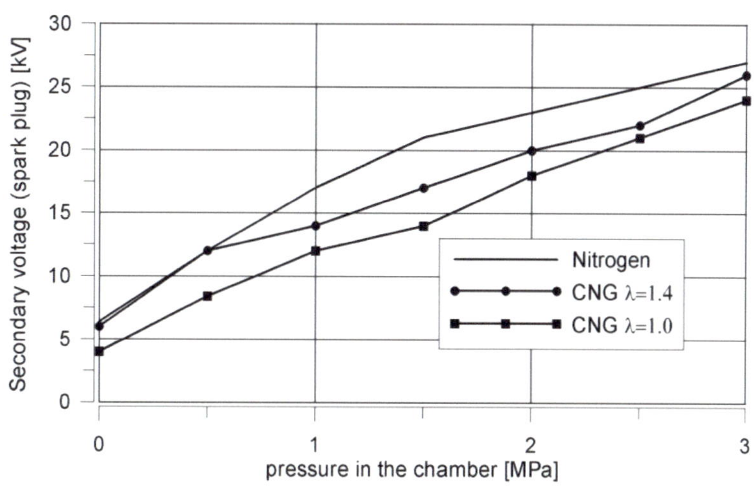

Figure 15. Influence of initial pressure on secondary voltage in ignition coil measured in caloric chamber filled by nitrogen and natural gas

7. Determination of energy losses during ignition

The model of ignition process takes into account only a small part of the spark plug and is shown in Figure 16.

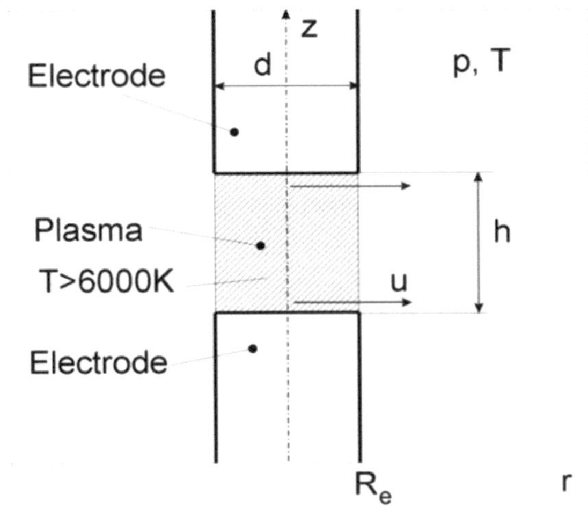

Figure 16. Model of spark ignition

During the sparking the plasma is formed between two electrodes and it is assumed to be smaller than the thickness of these electrodes. After short time a pressure shock takes place and the charge is moving on outer side with high velocity [1] [13]. The energy delivered directly to the charge is very low and therefore the energy losses should be assessed. As the experimental test showed, only a small part of delivered energy is consumed to increase the internal energy of the charge (maximum 10%). The energy losses during the ignition process can be divided into several kinds: radiation, breakdown, heat exchange with electrodes, kinetic energy which causes the turbulence, electromagnetic waves, flash and others.

7.1. Radiation energy of ignition

The part of the spark energy is consumed by radiation of the plasma kernel. The temperature T of plasma between two electrodes is above 6000 K. At assumption of the Boltzman radiation constant k=5.67 W/(m^2 K^4) and the coefficient of emissivity ε of a grey substance [9] for the ignition arc, the specific heat radiation e can be obtained from the formula:

$$e = \varepsilon \cdot k \cdot \left(\frac{T}{100} \right)^4 \tag{35}$$

The emissivity of the light grey substance was defined by Ramos and Flyn [4] and they amounted it in the range of 0.2 – 0.4. For that case it was assumed that ε = 0.3. The total radiation energy is a function of the ignition core surface A_i and sparking time t_i:

$$dE_r = A_i \cdot de \cdot dt = A_i \cdot \varepsilon \cdot k \cdot \left(\frac{T}{100}\right)^4 \cdot dt \tag{36}$$

At assumption that the temperature T of the arc increases proportionally with time from 6000 K to 300 K the total radiation energy can be calculated as follows:

$$E_r = A_i \cdot \varepsilon \cdot k \cdot \int_0^{t_i} \left(\frac{T}{100}\right)^4 dt = A_i \cdot \varepsilon \cdot k \cdot \left(\int_0^{t_i} \left(\frac{(T_{max} - T_2) \cdot t}{t_i \cdot 100}\right)^4 \cdot dt + \int_0^{t_i} \left(\frac{(T_{max} - T_2) \cdot t}{t_i \cdot 100}\right)^4 \cdot dt \right) \tag{37}$$

Assuming the radial shape of the core equal the radius $d/2$ of the electrodes and its height h equal the gap of electrodes and also that maximum temperature of the arc amounts 6000 K after t_1=20μs and then decreases to 800 K after t_2=2 ms, we can calculate the part of the coil energy as a loss of the radiation energy. Because 20 μs is comparably small with 2 ms then the equation (14) can be rewritten as follows:

$$E_r = \frac{A_i \cdot \varepsilon \cdot k \cdot (T_{max} - T_2)}{10^8} \cdot \frac{t_i}{5} \tag{38}$$

where A_i - the surface of the plasma core amounts $A_i = \pi \cdot d \cdot h$.

7.2. Ionization energy

Our experiment was carried out in nitrogen and on the basis of the literature data there are three ionization energies [7]: e_{i1} = 1402.3 kJ/mol, e_{i2} = 2856.0 kJ/mol, e_{i3} = 4578.0 kJ/mol. The energy required to breakdown of the spark is an ionization energy that can form later the arc. Total ionization energy can be calculated for n moles of the gas (nitrogen) in the core of plasma as:

$$E_i = n \cdot e_i = e_i \cdot \frac{p \cdot V_i}{(MR) \cdot T} = e_i \cdot \frac{\pi \cdot d^2 \cdot h}{4 \cdot (MR)} \cdot \frac{p}{T} \tag{39}$$

The initial temperature T amounts 300 K and universal gas constant (MR) = 8314 J/mol. For higher pressure, proportionally the higher ionization energy is required and the same is for lower temperature. However the plasma is formed with smaller radius, the ionization takes place in a higher volume with radius two times bigger.

7.3. Heat transfer to electrodes

A certain part of the energy delivered by the secondary circuit is consumed on the heating of the electrodes. In a small time of the sparking the heat transfer takes place on the small area approximately equal the cross section of the electrodes with diameter d. The main target is to determine the specific heat conductivity α between the gas and metal. This value α can be obtain from the Nusselt number Nu [2], gas conductivity λ_p and a characteristic flow dimension, in this case the diameter of the electrode:

$$\alpha = \frac{\mathrm{Nu} \cdot \lambda_p}{d} \tag{40}$$

where Nu is obtained from Reynolds number Re and Prandtl number Pr. However Ballal and Lefebvre [6] accounted for heat transfer the following expression for Nusselt number:

$$\mathrm{Nu} = 0{,}61 \cdot \mathrm{Re}_g^{0{,}46} = 0{,}61 \cdot \left(\frac{u_\infty \cdot d}{\mu} \right)^{0{,}46} \tag{41}$$

where u_∞ is gas velocity along the wall and μ is kinematic viscosity of gas. On the other hand the kinematic viscosity of the gas depends on the temperature T and density ρ according to the formula:

$$\mu = 5{,}18 \cdot 10^7 \cdot \frac{T^{0{,}62}}{\rho} \qquad \left[\mathrm{m}^2 / \mathrm{s} \right] \tag{42}$$

The conductivity of the gas is calculated based on the basis of Woschni [3] formula:

$$\lambda_p = 3{,}654 \cdot 10^{-4} \cdot T^{0{,}748} \qquad \left[\mathrm{W} / \left(\mathrm{m\ K} \right) \right] \tag{43}$$

Finally the cooling energy is calculated from the equation:

$$E_h = \left(\frac{1}{2} \pi \cdot d^2 \right) \cdot \alpha \cdot (T - T_w) \cdot t_i \tag{44}$$

7.4. Kinetic energy

Liu et al. [9] assumed that some fraction of the input energy is converted into kinetic energy of the turbulence according to the following formula:

$$\dot{E}_k = 4 \cdot \pi \cdot \left(\frac{d}{2} \right)^2 \cdot \rho_u \cdot u^3 \tag{45}$$

where ρ_u is density , u is the entrainment velocity and d is the kernel diameter. Using this equation the kinetic energy can be calculated for given values: $\rho_u = 1{,}403$ kg/m³ and for wave pressure moving with mean velocity u [m/s]. During ignition time t_i (less than 2 ms) the total kinetic energy amounts:

$$E_k = \int_0^{t_i} \dot{E}_k \mathrm{d}t = \dot{E}_k \cdot t_i \tag{46}$$

8. Ignition efficiency

Electric efficiency of the ignition systems define also the thermal resistance of these devices, because lower efficiency value decides about higher heating of the coil body and takes effect on their durability. On the basis of conducted tests by measurements of the primary (state 1) and secondary (state 2) current and voltage, it is possible to calculate the total electric efficiency of the ignition systems. The total electric efficiency can be defined as follows:

$$\eta_e = \frac{E_2}{E_1} = \frac{\int_{t_{02}}^{t_2} U_2 I_2 dt}{\int_{t_{01}}^{t_1} U_1 I_1 dt} \tag{47}$$

The electric efficiency for the ignition system with transistor ignition coil from Beru No 0040102002 is shown in Figure 17. The test of energy efficiency was done for 6 probes for every point of measurements. The electric efficiency is very small and at assumed initial pressures does not exceed 30%. The rest of energy goes into the surroundings in a form of heat. Lower efficiency is observed for nitrogen as the neutral gas. The same input energy for all considered cases amounted 210.74 mJ.

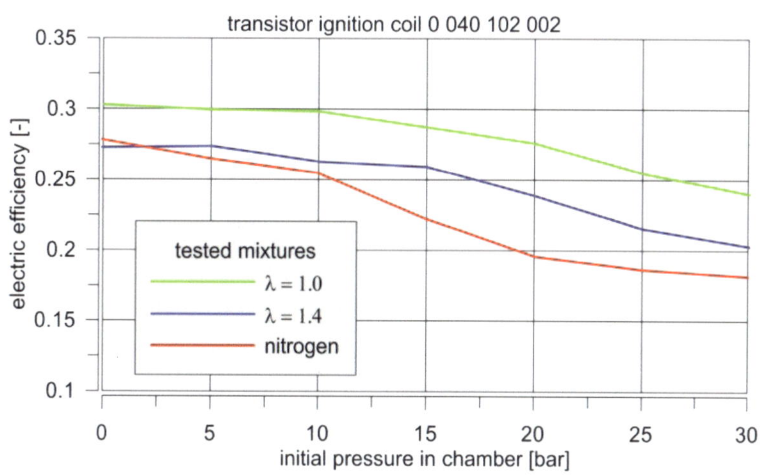

Figure 17. Electric efficiency of ignition system for two mixtures and nitrogen

9. Energy balance during ignition

On the basis of the carried out experimental tests and the theoretical considerations the balance of the energies delivered to the chamber from the secondary circuit of the coil can be done by Sankey chart. The carried out calculations determine the following values of heat losses for the case p = 25 bars and spark plug with the normal electrodes: 1) radiation - E_r = 7.8 mJ, 2) ionization - E_i = 7.2 mJ, 3) heat transfer - E_h = 31 mJ, 4) kinetic energy - E_k = 9 mJ.

Calculated total losses amount 55 mJ and measurements show that the thermal energy delivered to the charge E_{th} amounts only 4.23 mJ. On the other hand the measured energy delivered by the secondary circuit amounts E_2 = 61.05 mJ. The other non-considered heat losses amount E_c = 1.82 mJ. The graphical presentation of the participation of particular energies for the spark plug with normal electrodes and with 'thin' electrodes is shown on the Sankey diagram (Figure 18).

The energetic balance shows that the heat transfer to the electrodes consumes a half of delivered energy during the sparking process. Decrease of the cross-section of the electrodes to 25% of their initial value causes the increase of the thermal efficiency almost twice with decrease of the heat transfer to the electrodes. The work done by Liu et al [5] shows the discharge efficiency of different ignition system and for conventional spark ignition system this efficiency is below 0.1 (10%) despite the bigger coil energy (above 100 mJ).

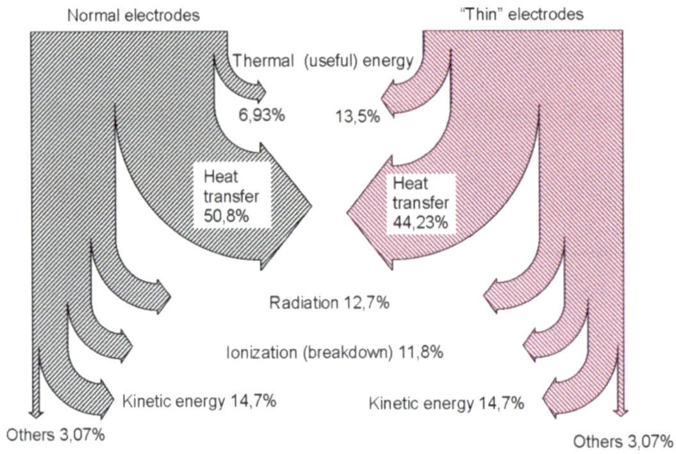

Figure 18. Balance of energy in the conventional ignition system for 2 types of the electrodes

10. CFD simulation of ignition and combustion process of CNG mixtures

Propagation of flame (temperature and gas velocity) depends on the temporary gas motion near the spark plug. The ignition process in SI gaseous engines was simulated in CFD programs (KIVA and Phoenics). Setting of the electrodes in direction of gas motion influences on spreading of the flame in the combustion chamber.

10.1. Propagation of ignition kernel

The propagation of the temperature during ignition process depends on the gas velocity between the spark electrodes. The experimental tests show an absence of the combustion process in the engine without gas motion. The combustion process can be extended with a big amount of hydrocarbons in the exhaust gases. The propagation of the temperature near the spark electrodes was simulated by use of Phoenics code for horizontal gas velocity amounted 10 m/s with taking into account the heat exchange, radiation, ionization and increase of the internal energy. The model of the spark ignition contained 40x40x1 cells with two solid blocks as electrodes and one block of the plasma kernel. The electrodes were heated during 1 ms with energy equal 8 mJ as it was determined during experimental tests. Propagation of the temperature near spark electrodes is shown in Figure 20 for two times 0.4 and 0.8 ms, respectively.

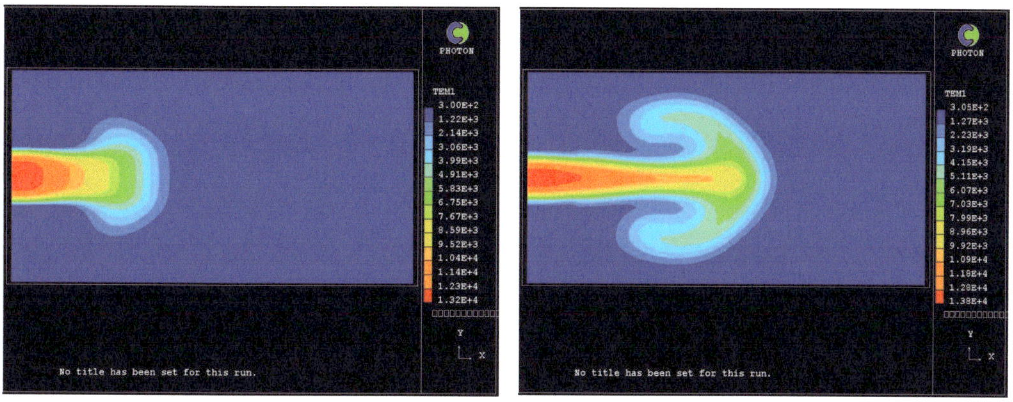

Figure 19. Temperature in the charge during ignition after 0.4 and 0.8 ms

The temperature inside the plasma grows as a function of the power of the secondary circuit in the coil and the velocity of the charge causes propagation of the temperature from the sparking arc outside of the plasma. Temperature inside the plasma kernel reaches value about 13000 K.

10.2. CNG ignition process in caloric chamber

The first step of the experimental tests was an observation of the ignition of the mixture of CNG and the air in the caloric chamber and the second step by use the simulation. The cylinder model has diameter D=34 mm and height B=22 mm. Volume of the chamber corresponds to the minimal volume of the combustion chamber in the engines of displacement 260 cm^3 and compression ratio 14.

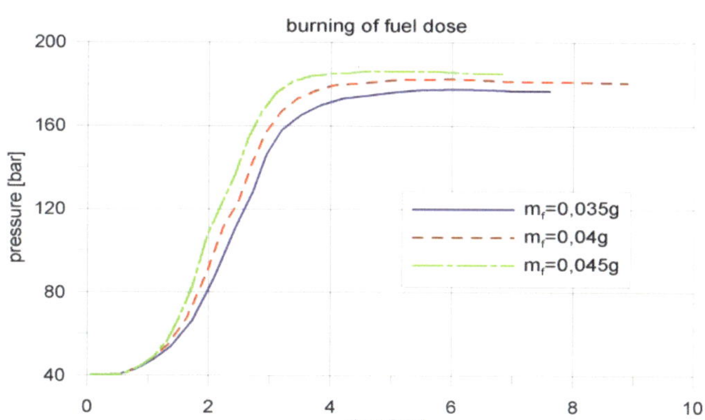

Figure 20. Increment of the pressure during combustion in the caloric chamber

Prediction of the mixture parameters in the chamber during combustion process was carried out by using the open source code of KIVA3V [4]. The complex test was conducted for 3 dose of CNG: 0.035, 0.04 and 0.045g, which corresponds to air excess coefficients λ: 1.58, 1.38 and 1.23, respectively at initial pressure 40 bars and temperature 600 K. At assumption of the high compression pressure in the caloric chamber it was obtained very high level of final pressure (about 180 bars) after burning of the whole dose (Figure 20). Velocity of increment of the mean charge temperature inside the caloric chamber depends on the value of the fuel dose (Figure 21) and for bigger dose the quicker increment of the temperature is observed.

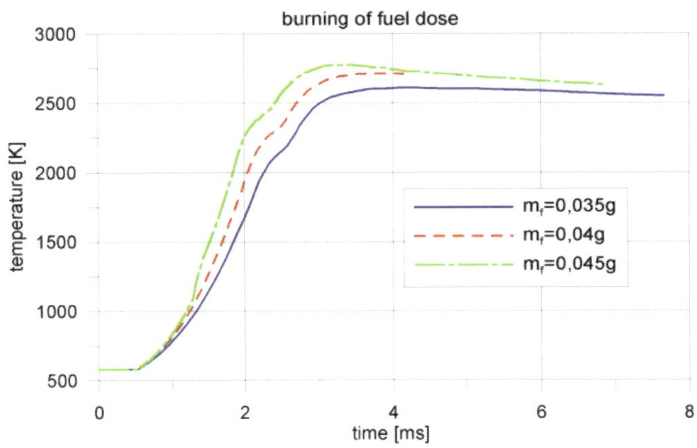

Figure 21. Variation of the temperature in the caloric chamber for different dose of CNG

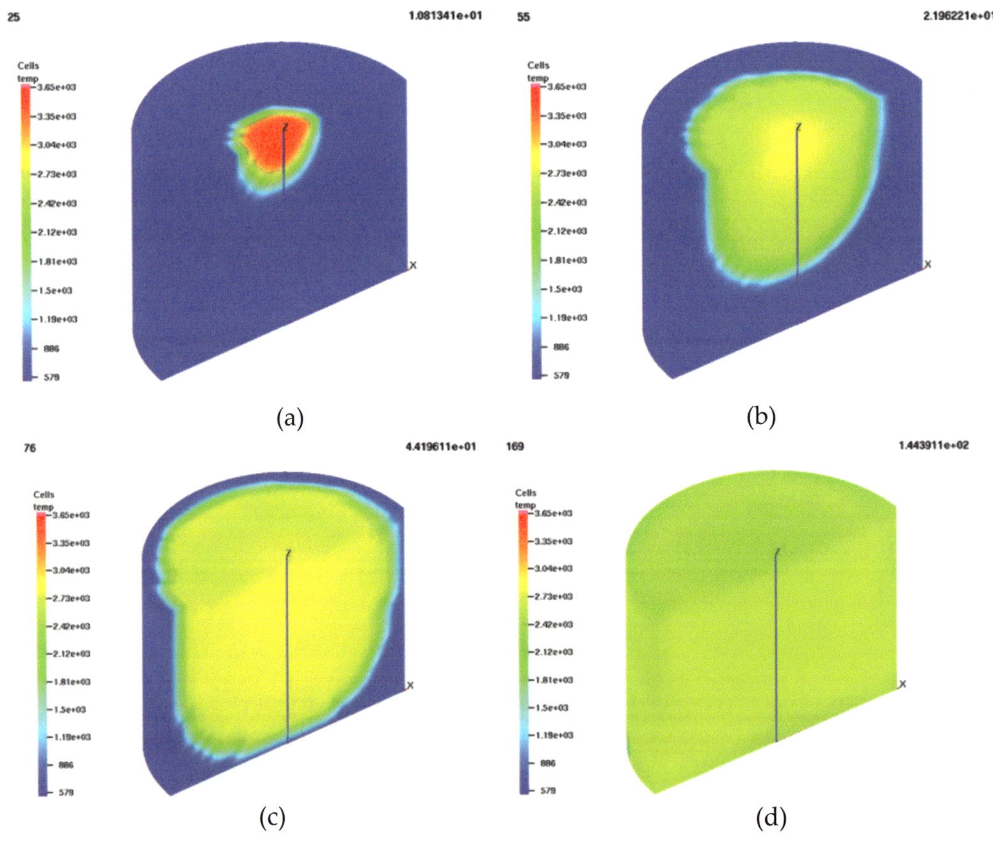

Figure 22. Temperature in the caloric chamber after initiation of combustion: a) 0.5 ms, b) 0.6 ms, c) 1.85 ms, d) 7.4 ms

The dose of fuel influences on variation of all thermodynamic parameters. The initiation of the combustion process lasted about 0.5 ms for all dose of the fuel. The complete combustion of all doses of the fuel without swirl and tumble follows after 4 ms with assumption of heat transfer to the walls. Four slides in Figure 22 show the spreading of the flame in the caloric chamber from the spark plug to the walls almost spherically. The maximum of temperature near spark plug amounts almost 3600 K and after combustion process decreases to 2700 K.

10.3. Verification of ignition modelling

The initial simulations of the CNG combustion was carried out on the model of the chamber used for the experimental tests on the Schlieren stand in a steady state initial conditions. The chamber had the volume equalled 100 cm^3 with diameter D=80 mm and width B=20 mm. The initiation of the ignition followed in the centre of the chamber by two thin electrodes. The chamber was filled by natural gas at 5 bars and λ=1.4. The initial temperature of the charge amounted 300 K, so this required much more electrical energy than for firing engine. The ignition energy was simulated as additional internal energy in the centre of the combustion chamber. The LES model for fully premixed charge was used in the CFD open source program OpenFOAM. The classical idea is to use a filter which allows for the separation of large and small length scales in the flow-field. Applying the filtering operator to the Navier-Stokes equations provides a new equation governing the large scales except for one term involving the small velocity scale. The model of combustion chamber was created by hexahedron cells and contained 68x68x32 cells. Calculations of combustion process were carried out in 64-bit Linux system with visualisation of results by use Paraview software. The combustion process in the chamber lasted a long time (above 50 ms), because of absence of the gas motion. The oxidation of methane was simulated by the OpenFOAM combustion procedure in Xoodles module. Thermodynamic properties of the charge were calculated by using JANAF tables. Increase of pressure in the flat combustion chamber without initial swirl or "tumble" of the charge is shown in Figure 23.

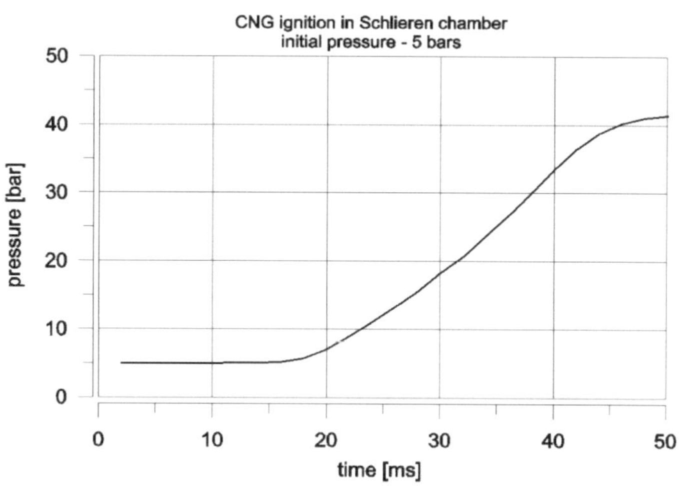

Figure 23. Increase of pressure in the chamber after ignition

The combustion process involves the change of thermodynamic parameters of the gas, which can be observed by moving of flame with different temperature, pressure and density in burned and unburned spaces. Full combustion of the methane-air mixture lasts longer than in the real engine combustion chamber at the same geometry of the combustion chamber. The propagation of chemical reactions is radial and the thick boundary of the combustion (about 8 mm) is observed because of the lean mixture. Propagation of the flame causes the radial compression of the gas between unburned and burned regions and thin area of twice higher density is formed. Figure 9 shows distribution of gas density in the chamber after 18 ms from start of ignition. Red colour indicates density on the level 0.0118 g/cm^3 and blue colour only 0.005 g/cm^3.

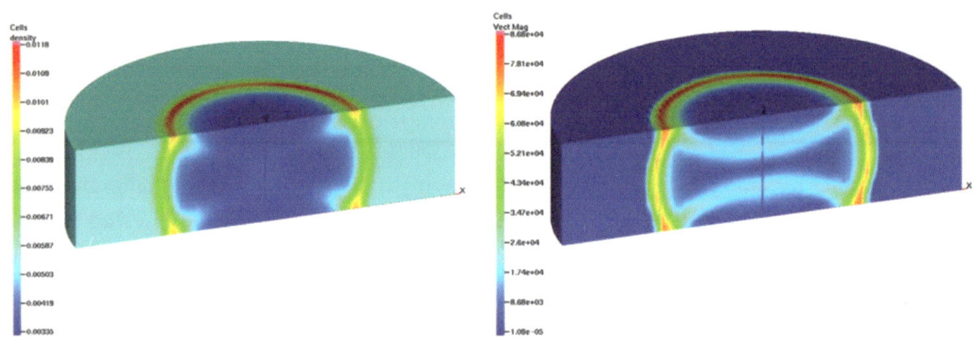

Figure 24. Gas density and absolute gas velocity after 18 ms from beginning of ignition

Combustion process in the narrow area takes place with turbulent velocity. Turbulence causes penetration of the flame into the unburned mixture with velocity higher than laminar combustion speed. For the methane-air stoichiometric mixture the combustion laminar speed amounts only 40 cm/s. For the considered case the absolute velocity of combustion in the flame region amounts about 80 m/s as one is shown in Fig.10. However, total combustion speed is very low and is close to the laminar speed of methane-air mixture 0.4 m/s.

Experimental tests on the Schlieren stand done by Sendyka and Noga [11] showed also radial propagation of the flame defined by the change of the charge density. Figure 25 shows the films of the flame propagation in the chamber at 3, 7, 40 and 54 ms after start of the ignition, respectively. The ignition of the CNG and air mixture with initial pressure 5 bars and initial temperature 300 K was initiated by two thin electrodes in centre of the combustion chamber. The charge was fully premixed with air excess ratio λ=1.4. The flame is distorted by touching into the quartz glass in the chamber, which is observed by hell circle inside the black circle. The change of gas density influences on the distortion of the laser beam and photos show development of the flame during combustion process. The experimental test proves the result obtained from simulation by using LES combustion model in the OpenFOAM program.

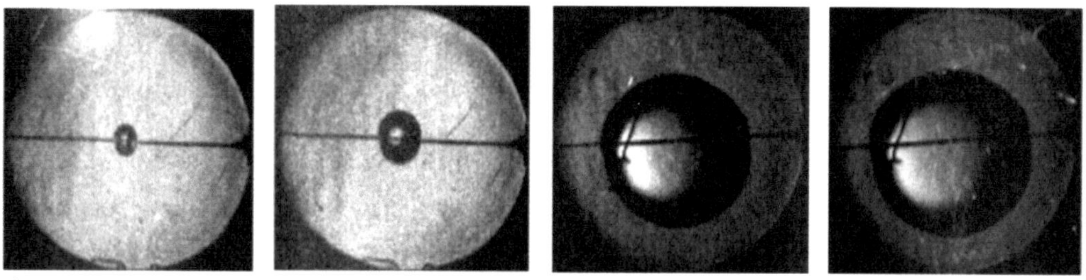

Figure 25. Schlieren stand – combustion boundary of the flame after 3, 7, 40 and 54 ms [15]

Both simulation and experiment do not show deviation of the spherical combustion flame. The experiment demonstrated velocity of combustion in radial direction of value 40 cm/s.

10.4. Mixture motion and ignition

The most important factor influencing on the ignition is the charge motion through the spark plug. Two kinds of motions were considered: swirl and tumble caused by valve and inlet profile, combustion chamber and squish. The combustion process is strongly connected with turbulence of the charge and only small part is the laminar speed of the total combustion velocity. Simulation was carried out in the rectangular space with central location of the spark plug. The mesh of the combustion chamber model with length and width 5 cm and height 3 cm was divided into 288000 cells with rectangular prism (NX=80, NY=80 and NZ=45). The calculations were carried out in transient conditions (initial time step 1e-6 s in time t=5 ms). The spark plug was located in the centre of the calculation space and the object of the electrodes was created by CAD system. The mesh in the region of the spark plug electrodes contains fine grids with cell length equal 0.3 mm in x and y axis.

At the first the ignition of CNG was simulated with „initial tumble" ω_y =250 rad/s and p=20 bars. The charge with velocity about 15 m/s flew through the gap of the spark plug causing the propagation of the flame inside the chamber. The simulation of combustion and gas movement was carried out also by Phoenics, which takes into account turbulence model and simple combustion of compressible fluid. The charge motion is connected with high turbulence and this causes also the higher combustion rate.

Distribution of the combustion products in the modelled space is shown in Figure 26 at 0.5 ms and 1.2 ms after start of the ignition, respectively. After short time (about 1 ms) the whole charge is burned in the calculation space. The higher flow velocity is between the electrodes of the spark plug. The other simulation was carried out for the central swirl around the spark plug with swirl velocity 15 m/s on the mean radius 1.5 cm. In this case the interaction of the electrode shape is seen – the propagation of the flame is faster in the opened site of the electrodes. Figure 27 presents development of combustion process after 1 and 4 ms from beginning of the ignition.

The swirl in the chamber influences on the irregular propagation on the flame and extends the combustion process. Even after 4 ms the combustion of the methane is not full. Velocity of the gas flow in the spark plug gap is smaller than in the "tumble" case. For this reason the propagation of the combustion products and flame is not uniform.

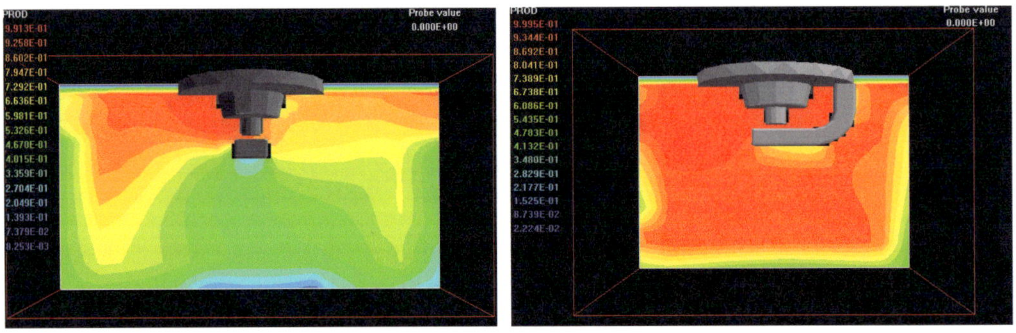

Figure 26. Combustion products with initial "tumble" charge motion after 0.5 ms and after 1.2 ms

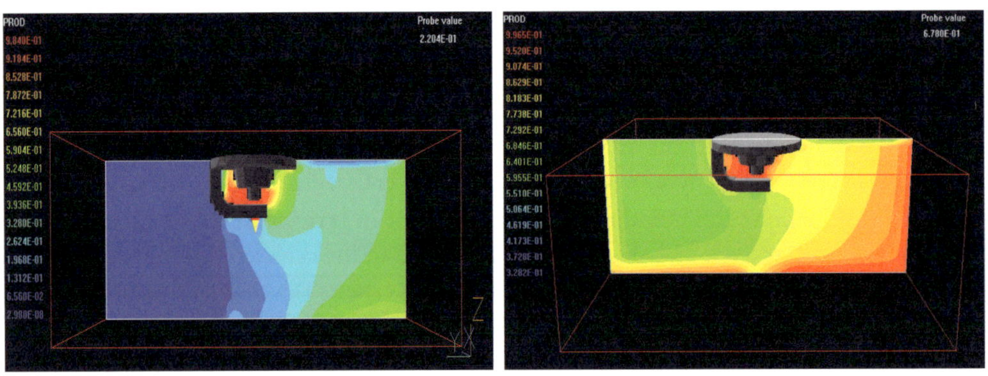

Figure 27. Combustion products with initial "swirl" charge motion after 1 ms and after 4 ms

11. Conclusions

The chapter contains results of theoretical, modelling and experimental work considered to factors, which have very big impact on the ignition of gaseous fuels in combustion engines. On the fact of more and more important role of gaseous engines, particularly those fuelled by natural gas, definition of good conditions for ignition of gaseous fuels is one of the task of development of modern spark ignition gaseous engines, particularly with high charging ratio. Experimental works with CNG ignition were done in the caloric chamber, however in conditions closed to real conditions of engine work. On the presented considerations one can draw some conclusions and remarks:

1. Gaseous fuels, such as CNG requires higher electric energy delivered by the ignition system. Higher pressure in the combustion chamber increases internal energy near the spark plug and requires also higher secondary voltage of the ignition coil. For gaseous leaner mixtures an ignition system with higher energy is needed (above 60 mJ).

2. The higher initial pressure increases the thermal efficiency of the ignition system.
3. For the conventional ignition systems even with high secondary energy above 60 mJ only a small part maximum 15% of it is consumed by the charge.
4. The maximum of thermal efficiency was obtained at initial pressure 25 bars with value of 13, 5% for the spark plug with thin electrodes and only 1% at ambient pressure and temperature.
5. The spark plug with thin electrodes indicates higher thermal efficiency than the spark plug with normal electrodes. This is caused by small heat exchange with the electrode' walls.
6. The energy losses consist of heat exchange, ionization energy (breakdown), radiation and others. The biggest of them are the heat transfer to the spark electrodes and radiation.
7. On the basis of CFD simulation one proved that nature of mixture motion (tumble or swirl) in the combustion chamber influences on propagation velocity of the ignition kernel and combustion process.
8. Ignition in CNG diesel engine can be caused by injection of small ignition dose of diesel oil.

Author details

Wladyslaw Mitianiec
Cracow University of Technology, Cracow, Poland

Acknowledgement

The author would like to acknowledge the financial support of the European Commission in integrated project "NICE" (contract TIP3-CT-2004-506201). Personally I would like to acknowledge the contribution of the Cracow University of Technology, particularly Prof. B. Sendyka for supervising during the project and Dr. M. Noga for his big input in the experimental work.

12. References

[1] Heywood J (1988), Internal Combustion Engine Fundamentals, Mc Graw-Hill, New York
[2] Look D.C., Sauer H.J (1986), Engineering Thermodynamics, PWS Engineering, Boston
[3] Mitianiec W., Jaroszewski A. (1993), Mathematical models of physical processes in small power combustion engines (in polish), Ossolineum, Wroclaw-Warszawa-Krakow
[4] Ramos J.I (1989), Internal combustion modeling, Hemisphere Publishing Corporation, New York
[5] Amsden A.A. et al (1989), KIVA-II - A Computer Program for Chemically Reactive Flows with Sprays, Los Alamos National Lab., LA-11560-MS

[6] Ballal D., Lefebvre A (1981), The Influence of Flow Parameters on Minimum Ignition Energy and Quenching Distance, 15th Symposium on Combustion, pp.1737-1746, The Combustion Institute, Pittsburgh
[7] Eriksson L (1999), Spark Advance Modeling and Control, Linkoping University, dissertation No 580, Linkoping
[8] Hires S.D., Tabaczyński R.J (1978), The Prediction of Ignition Delay and Combustion Intervals for Homogeneous Charge, Spark Ignition Engine, SAE Pap. 780232
[9] Liu J.,Wang F., Lee L., Theiss N., Ronney P (2004), Gundersen M., Effect of Discharge Energy and cavity Geometry on Flame Ignition by Transient Plasma, 42nd Aerospace Sciences Meeting, 6th Weakly Ionized Gases Workshop, Reno, Nevada
[10] Maly R., Vogel M (1979), Initiation and propagation of Flame Fronts in Lean CH4 – Air Mixtures by a Three Modes of the Ignition Spark, Seventeenth Symphosium on Combustion, pp 821-831, The Combustion Institute, Pittsburgh
[11] Sendyka B., Noga M (2007), Propagation of flame whirl at combustion of lean natural gas charge in a chamber of cylindrical shape, Combustion Engines, 2007-SC2, Bielsko Biała
[12] Spadaccini L.J.,Tevelde J.A (1980), Autoignition Characteristic of Aircraft Fuels, NASA Contractor Report Cr-159886
[13] Thiele M., Selle S., Riedel U., Warnatz J.,Maas U (2000), Numerical simulation of spark ignition including ionization, Proceedings of the Combustion Institute, Volume 28, pp. 1177- 1185
[14] Vandebroek L.,Winter H., Berghmans J (2000), Numerical Study of the Auto-ignition Process in Gas Mixtures using Chemical Kinetics, K.U.Leuven, Dept. Of Mechanical Engineering, 2000
[15] FEV (2004) information materials, Aachen

Fundamental Studies on the Chemical Changes and Its Combustion Properties of Hydrocarbon Compounds by Ozone Injection

Yoshihito Yagyu, Hideo Nagata, Nobuya Hayashi, Hiroharu Kawasaki, Tamiko Ohshima, Yoshiaki Suda and Seiji Baba

Additional information is available at the end of the chapter

http://dx.doi.org/10.5772/54706

1. Introduction

The prevention of global warming caused by increasing of CO_2 has been high interest to the public and grows environmental awareness, and also effective use of fossil fuel especially for a vehicle with internal combustion engine, such as an automobile, a ship and an aircraft, has been developed in several research field. Hydrogen for fuel cell, bio-diesel fuel (BDF), dimethyl ether (DME) and ethanol as alternative energy source for vehicle has been widely studied (S. Verhelst&T. Wallner, 2009, M.Y. Kim et al., 2008, B. Kegl, 2011, T.M.I. Mahlia et al., 2012) and some of them have begun to exist at the market in the recent decade. However, these new energy sources may take a long time until reaching at a real general use because of high investment not only for improving an infrastructure but retrofitting a system of vehicle.

Effects of ozone addition on combustion of fuels were reported as original research works. Combustion of natural gas with ozone was studied, and reduction of concentration of CO and C_nH_m by ozone injection was observed (M. Wilk&A. Magdziarz, 2010). Ozone injection to internal compression engine is effective to decrease in the emission rate of NO, CO and CH and an improvement of cetane number and fuel efficiency (T. Tachibana et al., 1997). Also there are patent applications covering the invention developed by the effect of ozone injection (Imagineering Inc.: JP 19-113570(A), Sun chemical Co., Ltd.: JP 7-301160, Nissan Motor Co., Ltd.: JP 15-113570(A)). Although the results of ozone injection were often reported, the mechanism of the phenomenon is still unclear. Discharge system for generating ozone was usually introduced before or directly in a cylinder, and then before expansion stroke the excited oxygen species including ozone reacts on hydrocarbon compounds, such as petrol and light diesel oil. Previous reports are chiefly focused on the phenomena/effects and are not mentioned. We focused on the measurement of reaction products and investigated chemical changes among vapourised hydrocarbon compounds and activated dry air or oxygen including ozone by exposing discharge in this paper.

2. Material and method

2.1. Discharge reactor

Discharge was generated on the surface of a grid pattern discharge element (size: 85mm × 40mm, grid size: 5mm with 1mm in width of electrode) placed in the discharge reactor. The element was designed for generating surface discharge, and the discharge voltage was generated by high AC voltage power supply (Logy Electric Co., LHV-13AC) and varying from 8.0kV to 13.0kV (with approx. 9kHz~11kHz in frequency). Ozone concentration was measured as a standard of active oxygen species and was generated between 0.62g/m^3 to 8.8g/m^3 by dry air or pure oxygen.

2.2. Reactor vessel and hydrocarbon compounds

Petrol and light diesel oil were used for reaction with dry air or oxygen exposed to discharge. We used commercially available petrol and light diesel oil in the test, and the octane number of petrol and the cetane number of light diesel oil are among 90 to 92 and among 53 to 55 respectively. Hydrocarbon compounds as typified by petrol and light diesel oil consist of a lot of chemical substances. Isooctane (2.2.4-trimethylpentane, C_8H_{18}), which are one of the components of petrol and light diesel oil, are also applied as reacting substance of the test. These hydrocarbon compounds were vapourized in the reactor vessel (Figure 1), and air or oxygen exposed to discharge was injected through inlet to investigate chemical reaction between active species and vapourised hydrocarbon compounds. The condition of the entire experiment was not controlled and was carried out under an atmospheric pressure and temperature around 23 degrees Celsius.

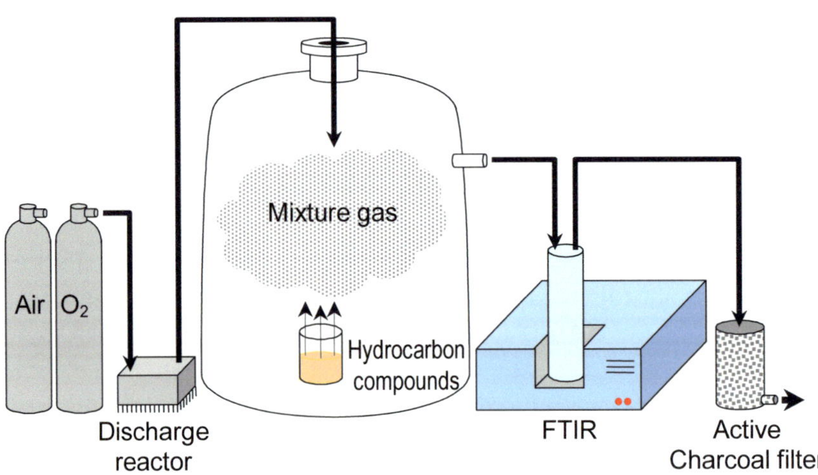

Figure 1. A schematic of the experimental apparatus

2.3. Analysis of the mixture gas

A composition of the mixture gas of dry air exposed to discharge and vapourised hydrocarbon compounds was detected by Fourier transform infrared spectroscopy (FTIR, Shimadzu corp., FTIR-8900) and gas chromatography mass spectroscopy analysis (GC-MS, Shimadzu corp., GCMS-QP2010). Twenty-two meters reflective long-path distance gas cell (Pike Technologies inc., Permanently Aligned Gas Cell 162-2530) was installed to the FTIR for detecting the detail of the gas. Wavenumber of the mixture gas was measured between $4000cm^{-1}$ to $1000cm^{-1}$ and the FTIR spectra of the mixture gas were accumulated to 40 times.

Temperature of the vapourizing chamber of GC (gas chromatography) was set up from 30 to 250 degrees Celsius with increasing 10 degrees Celsius a minute, and entire measurement time was 24 minutes. Column flow stayed constant on 2.43mL a minute and column pressure was kept 97.9kPa. Mass-to-charge ratio in MS (mass spectrometry) was detected from 35 to 250m/z.

3. FTIR analysis

First of all of the investigations, dry air exposed to discharge and hydrocarbon compounds was mixed in the transparent bell jar (approximately 18L). Fifty mL of commercially available petrol (octan number 90 ~ 92) was poured into the beaker, and it was placed in the centre of the bell jar. Flow rate of the injected gas to the bell jar was consistently kept on 12L/min. The vapourized hydrocarbon compounds became clouded immediately after injecting dry air exposed to the discharge, and it grew into a dense fume increasingly (Figure 2). It was suggested that active species in the gas decomposed a part of structure of hydrocarbon compounds and water molecules were generated. Thus the generation of cloudy mixture gas was probably caused by saturated water vapour.

Figure 2. A picture of the bell jar with discharge (left) and without discharge (right).

Isooctane (2.2.4-trimethylpentane, C_8H_{18}) as elemental substances of petrol was mixed with dry air through the discharge chamber. Ozone concentration was adopted as a measure of several parameters and was kept at 2.7g/m^3 in the tests. FTIR spectra of the mixture gases were indicated in Figure 3. The peak at 1725cm^{-1} was remarkably detected at FTIR spectra. It was supposed to be a reaction product, which were generated from hydrocarbon compound and active species in the discharged dry air. Relatively-sharp feature near 1058cm^{-1} and 2120cm^{-1} of the FTIR spectra was due to extra ozone. Also CO_2 near 2360cm^{-1} and N_2O near 2237cm^{-1} were detected. Broad features of hydrocarbon compounds around 3000cm^{-1} particularly decreased after injecting discharged dry air. These disappearing or decreasing spectra such as around 3000cm^{-1} were changed into other chemical structure. However, H_2O was not observed in the spectra because it might be much weaker than other spectra. It was clearly found that the injection of discharged dry air caused partial chemical changes of isooctane from the FTIR spectra analysis.

We analysed the spectra of 1725cm^{-1} prominently figured in both FTIR spectra as reaction product. Oxidative products of hydrocarbon compounds were mainly searched and some candidates, which had strong peak near 1725cm^{-1}, were found such as heptanal ($C_7H_{14}O$), 2-hexanone ($C_7H_{14}O$), acetone (C_3H_6O), octanal ($C_8H_{16}O$), trans-cyclohexane ($C_8H_{16}O_2$) and dodecanal ($C_{12}H_{24}O$). FTIR spectra of octanal that was one of these hydrocarbon oxide/dioxide compounds were indicated in Figure 4. Ozone was possibly major element of generating oxidative products of hydrocarbon compounds because a lifetime of other active oxygen species, which were for instance oxygen atom, singlet oxygen molecule and hydroxide, was very shorter than it.

Typical FTIR spectra of the mixture gas consisting of vapourized petrol and dry air exposed to the discharge were indicated in Figure 4. Almost same spectra of Isooctane (2.2.4-trimethylpentane) (Figure 3) at around 1725cm^{-1} were also detected as the reaction product although these spectra were not observed before injecting dry air passed through the discharge reactor. The production rate of reaction products was gradually saturated with increasing ozone concentration (Figure 5). The production rate of the newly detected spectra and the decreasing peak of hydrocarbon compounds were strongly dependent on the concentration of ozone in the discharged dry air. Typical peaks of ozone gas were observed at 1058cm-1 and 2120cm^{-1} as extra ozone, which did not react to hydrocarbon compounds. The peaks of reaction products and the extra ozone gas were detected in the case that the thinnest 0.62g/m^3 of ozone was injected in this investigation. FTIR spectra above 3200cm^{-1} tend to indicate lower % of transmittance. Shorter wavelength of infrared light has tendency to be scattered about small particles in the long-path distance gas cell such as water molecule obtained in the reaction between ozone and hydrocarbon compound.

The spectra of vapourised light diesel oil blended with ozone were similar to other FTIR spectra (Figure 6). The peaks of reaction products at around 1725cm^{-1} were also found. The production rate of the reaction product near 1725cm^{-1} was less than the case of petrol because petrol indicated higher evapouration rate due to lower specific gravity in comparison with light diesel oil.

Ozone is one of the essential active gases for chemically changing the composition of vapourized hydrocarbon compounds. It was supposed that ozone decomposed CH band of hydrocarbon compounds partially and ozonised hydrocarbon compounds were produced as these reaction products at 1725cm^{-1}. In conclusion, it was found that the reaction products, which were also generated by the thinnest ozone concentration in the test, were made by the injected ozone and the hydrocarbon compounds. These ozonised hydrocarbon compounds might be directly/indirectly contributed toward the improvement more efficient combustion of the internal combustion engine.

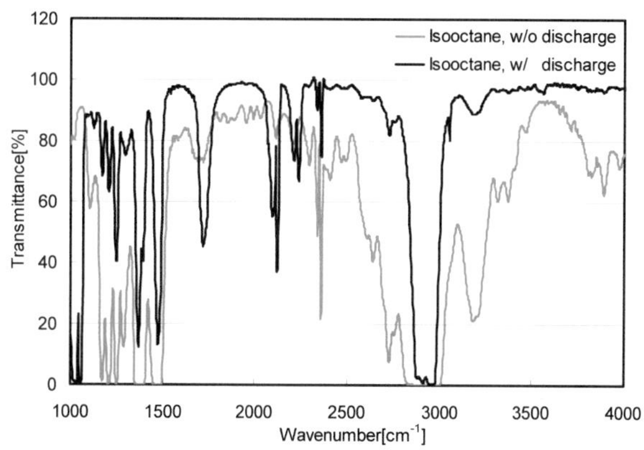

Figure 3. Typical FTIR spectra of the mixture gas blended isooctane (2.2.4-trimethylpentane, C_8H_{18}) and dry air exposed to discharge

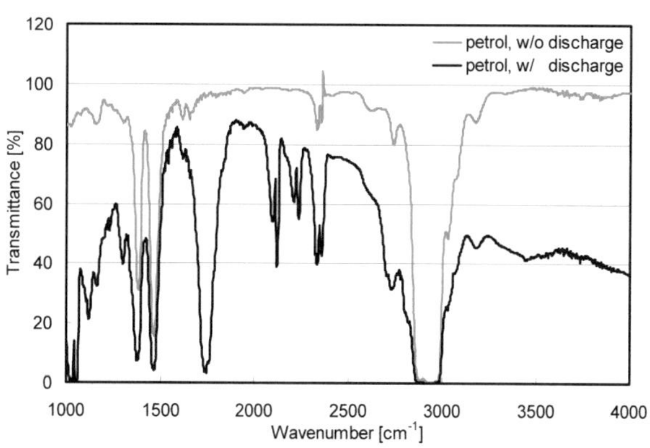

Figure 4. Typical FTIR spectra of the mixture gas blended petrol and dry air exposed to discharge

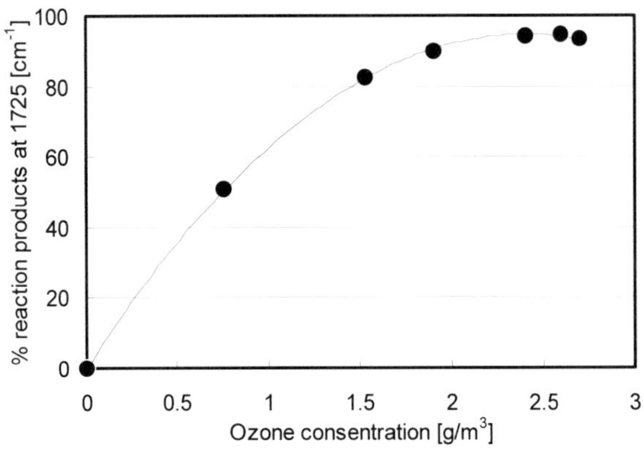

Figure 5. Generation rate of reaction products near 1725cm^{-1} of FTIR spectra

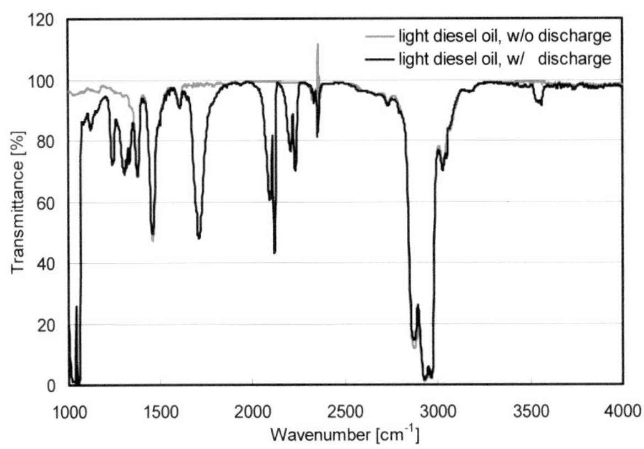

Figure 6. Typical FTIR spectra of the mixture gas blended light diesel oil and dry air exposed to discharge.

4. GC-MS analysis

The details of reaction products, especially to search the composition of the strong peak detected near 1725cm^{-1} by FTIR, were analysed by GC-MS. Petrol vapourised at room temperature (23 degrees Celsius) was injected to inlet of the GC-MS and its retention time data of GC were indicated in Figure 7. Also, vapourised petrol was blended in the closed container with oxygen passed through the discharge reactor. The mixture gas was introduced to the GC-MS directly and its spectra were indicated in Figure 8. Spectral lines of GC-MS were identified by spectral library and chemical substances were mainly indicated in Table 1. Chemical substances were searched when relative intensity obtained more than 5 arbitrary units. Shaded cells in column of "without Ozone" in Table 1 indicated chemically changed substances after ozone injection. Also, shaded cells in column of "with Ozone" of Table 1 indicated newly produced mainly oxidative products of hydrocarbon compounds by ozone injection.

It was found that after ozone injection hydrocarbon compounds were combined with oxygen atom or oxygen molecule. Vapourised petrol was comprised of approximately 300 kinds of hydrocarbon compounds, and several compounds with a high amount rate were detected from the vapourised petrol by GC-MS such as butane (C_4H_{10}) pentane (C_5H_{12}) and hexane (C_6H_{14}) (Table 1; without ozone). In the case after ozone injection, several reactive products were detected such as acetaldehyde (C_2H_4O), heptanal ($C_7H_{14}O$), 2-Hexanone ($C_7H_{14}O$) and octanal ($C_8H_{16}O$) (Table 1; with ozone).

Number of carbon atom after ozone injection had tendency to be higher than the vapourised petrol without ozone treatment. It was expected that decomposed hydrocarbon compounds recombined themselves after that ozone decomposed CH band partially.

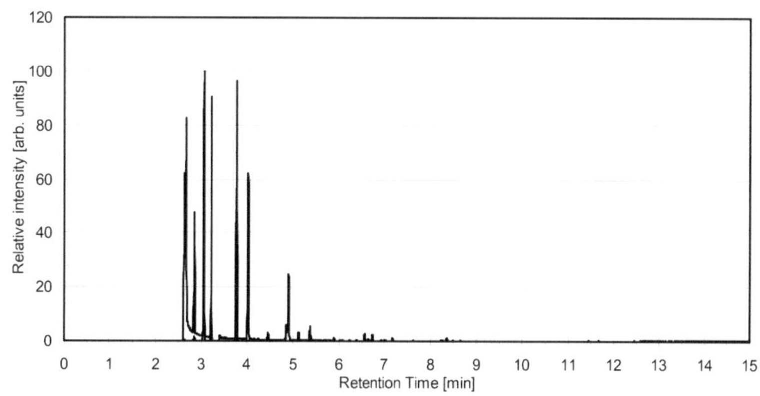

Figure 7. Typical retention time data of petrol without ozone.

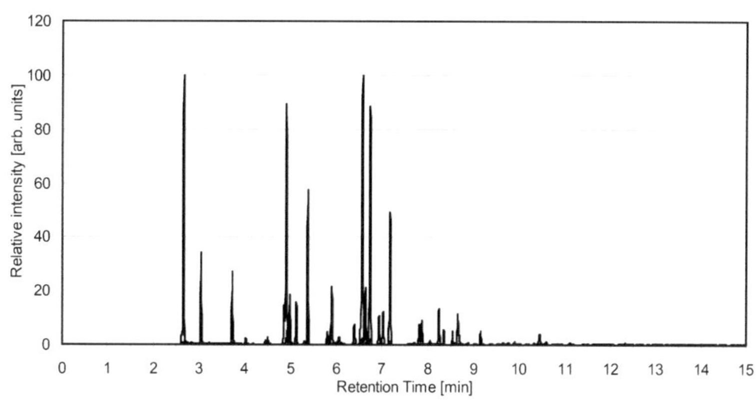

Figure 8. Typical retention time data of petrol with ozone.

The FTIR spectra of the reaction products, which were detected hydrocarbon compounds combined with oxygen atom or molecule, were searched by scientific database of The National Institute of Standards and Technology (NIST Standard Reference Database). It was found that many of reaction products had FTIR spectra near 1725cm^{-1} such as acetaldehyde (C_2H_4O), acetone (C_3H_6O), heptanal ($C_7H_{14}O$), octanal ($C_8H_{16}O$) and dodecanal ($C_{12}H_{24}O$)

(Table 1). As an example of them, FTIR spectra of Acetone (C_3H_6O) and octanal ($C_8H_{16}O$) were indicated in Figure 9 and Figure 10, respectively.

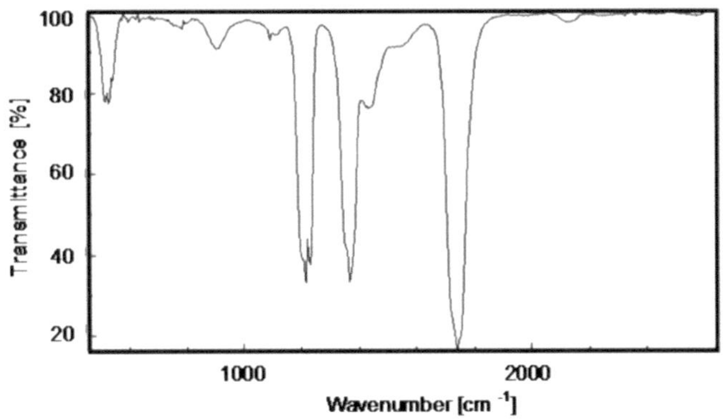

Figure 9. FTIR spectra of acetone (C_3H_6O), which is one of the oxidative products of hydrocarbon compounds (NIST Standard Reference Database).

without Ozone					with Ozone				
Ret. time	chemical formula	chemical substance	similarity	M.W.	Ret. time	chemical formula	chemical substance	similarity	M.W.
					2.66	$C_3H_2O_2$	Propiolic acid	98	70
					3.04	C_2H_4O	Acetaldehyde	99	44
3.06	C_4H_{10}	Isobutane	97	58					
3.22	C_4H_{10}	Butane	97	58					
					3.72	C_3H_6O	Acetone	97	58
3.77	C_5H_{12}	Butane, 2-methyl-	97	72					
4.02	C_5H_{12}	Pentane	98	72					
4.46	C_6H_{14}	Butane, 2.2-dimethyl-	96	86					
4.86	C_6H_{14}	Butane, 2.3-dimethyl-	94	86	4.86	C_6H_{14}	Butane, 2.3-dimethyl-	96	86
4.9	C_6H_{14}	Pentane, 2-methyl-	94	86	4.9	C_6H_{14}	Pentane, 2-methyl-	95	86
					4.99	C_4H_8O	2-Butanone	97	72
5.13	C_6H_{14}	Pentane, 3-methyl-	98	98	5.13	C_6H_{14}	Pentane, 3-methyl-	98	86
					5.13	C_6H_{14}	Pentane, 3-methyl-	98	86
5.38	C_6H_{14}	Hexane	98	86	5.38	C_6H_{14}	Hexane	98	86
					5.8	C_7H_{16}	Pentane, 2.2-dimethyl-	96	100
					5.89	C_6H_{12}	Cyclopentane, methyl-	93	84
					5.98	$C_5H_{10}O$	Butanal, 3-methyl-	95	86
					6.06	$C_5H_{10}O$	2-Pentanone	83	86
					6.39	C_7H_{16}	Pentane, 3.3-dimethyl-	93	100
6.57	C_7H_{16}	Heptane, 3-methyl-	92	114	6.57	C_7H_{16}	Heptane, 3-methyl-	93	100
					6.64	C_7H_{16}	Pentane, 2.3-dimethyl-	96	100
6.74	C_7H_{16}	Hexane, 3-methyl-	95	100	6.74	C_7H_{16}	Hexane, 3-methyl-	96	100
					6.94	$C_{10}H_{20}$	1-Octane, 3-ethyl-	91	140
					7.03	C_8H_{18}	Hexane, 2,2-dimethyl-	95	114
					7.18	C_7H_{16}	Hexane, 3-methyl-	94	100

without Ozone					with Ozone				
Ret. time	chemical formula	chemical substance	similarity	M.W.	Ret. time	chemical formula	chemical substance	similarity	M.W.
					7.82	C_8H_{18}	Hexane, 2,5-dimethyl-	95	114
					7.87	C_8H_{18}	Hexane, 2,4-dimethyl-	95	114
					8.24	C_8H_{18}	Pentane, 2,3,4-trimethyl-	98	114
					8.35	C_7H_8	Toluen	98	92
					8.54	C_8H_{16}	Heptane, 4-methyl-	96	114
					8.65	C_8H_{18}	Heptane, 3-methyl-	93	114
					8.83	C_8H_{16}	Cyclohexane, 1,3-dimethyl-, trans-	93	112
					8.87	C_9H_{20}	Hexane, 2,2,5-trimethyl-	90	128
					9.01	C_8H_{16}	Cyclopentane, 1-ethyl-3-methyl, trans-	94	112
					9.05	C_8H_{16}	Cyclopentane, 1-ethyl-3-methyl, trans-	94	112
					9.15	C_8H_{18}	Hexane, 2,4-dimethyl-	93	114
					9.38	C_8H_{16}	Cylohexane, 1,4-dimethyl-, cis-	93	112
					9.53	C_9H_{20}	Hexane, 2,3,5-trimethyl-	86	128
					9.65	C_8H_{18}	Heptane, 2,4,6-trimethyl-	90	142
					9.73	C_6HO_2	2,3-dioxabicyclo[2,2,1]heptane, 1-methyl-	82	114
					9.76	$C_{11}H_{24}$	Octane, 2,3,3-triethyl-	88	156
					9.9	C_9H_{20}	heptane, 2,5-dimethyl-	93	128
					9.98	$C_7H_{14}O$	Heptanal	82	114
					10	$C_5H_{10}O$	Butanal, 3-methyl-	78	86
					10.3	$C_{12}H_{26}$	Octane, 4,5-diethyl-	87	170
					10.4	C_8H_{10}	o-Xylene	94	106
					10.4	$C_{12}H_{18}$	Benzene, (3,3-dimethylbutyl)-	80	162
					10.5	$C_7H_{12}O$	2-Hexanone, 4-methyl-	83	114
					10.6	C_9H_{20}	Octane, 3-methyl-	93	128
					10.8	$C_8H_{16}O$	Octanal	83	128
					11.1	$C_{12}H_{26}O$	1-Octanol, 2-butyl-	89	186
					11.2	C_7H_6O	Heptanal	82	114
					11.4	$C_{17}H_{28}$	5-Tetradecane, (Z)-	82	196
					11.5	C_9H_{12}	Benzene, (1-methylethyl)-	83	120
					11.6	$C_{12}H_{26}O$	1-Decanol, 2-ethyl-	87	186
					11.7	$C_8H_{16}O$	Octanal	83	128
					11.8	$C_9H_{20}O$	1-Pentanol, 4-methyl-2-propyl-	85	144
					11.9	C_7H_6O	Benzaldehyde	78	106
					11.9	$C_{10}H_{20}O$	isooctane,(ethyenyloxy)-	81	156
					12	$C_{11}H_{17}N$	n-Butylbenzylamine	79	163
					12.2	C_9H_{12}	Benzene, 1-ethyl-3-methyl-	93	120
					12.3	$C_8H_{16}O$	Octanal	74	128
					12.3	$C_{12}H_{26}O$	1-Decanol, 2-ethyl-	76	186
					12.3	C_9H_{12}	Benzene, 1,3,5-trimethyl-	85	120
					12.5	$C_{12}H_{24}$	1-Decanol, 3,4-dimethyl-	87	168
					12.7	$C_8H_{16}O$	Octanal	86	128
					13	$C_{10}H_{20}O$	isooctane,(ethyenyloxy)-	84	156
					13.1	$C_8H_{16}O$	Octanal	79	128
					13.1	$C_8H_{16}O$	Octanal	75	128
					13.4	C_9H_{12}	Benzene	92	120
					13.4	$C_{12}H_{24}O$	Decanal	78	156
					13.5	$C_{10}H_{22}O$	1-Octanol, 2,7-dimetyl-	78	158
					13.6	$C_{24}H_{48}O_2$	Hexanoic acid, octadecy ester	75	368
					13.7	$C_8H_{16}O$	Octanal	74	128
					14	$C_{12}H_{24}O$	Dodecanal	79	184
					14	$C_{10}H_{20}O$	Decanal	80	156
					14.1	$C_{10}H_{20}O$	Decanal	78	156

without Ozone					with Ozone				
Ret. time	chemical formula	chemical substance	similarity	M.W.	Ret. time	chemical formula	chemical substance	similarity	M.W.
					14.2	$C_{11}H_{22}O$	Undecanal	80	170
					14.5	$C_9H_{18}O$	Nonanal	85	142
					14.7	$C_{12}H_{24}O$	Dodecanal	77	184
					16.1	$C_{10}H_{20}O$	Decanal	87	156

Table 1. List of chemical substances detected by GC-MS with and without ozone.

5. Combustion properties of n-Octane by ozone

5.1. Reactor and analysis

Vapour-phase combustion reactions of n-octane with ozone are tested in a flow reactor indicated in Figure 10. Flow reactor mainly consists of discharge reactor to make ozonized air, saturator vessel to vaporise n-octane, electric furnace to combust vaporised n-octane and measure combustion temperature. Sampling point is located at the end of the flow reactor to analyse combustion reactions properties in the electric furnace. Ozone in gaseous form is generated by discharge reactor, which has four surface discharge chips indicated in Figure 11. Dry air is a source of ozone generation at discharge reactor and ozone concentration is maintained at 6.4g/m^3 in the test.

Carbon dioxide, carbon monoxide, oxygen and nitrogen in the exhaust gas, which is generated in the vapour-phase combustion reactions of n-octane, are analysed by gas chromatography with thermal conductivity detector (GC-TCD). Organic compounds in a gas at sampling point are analysed by gas chromatography with flame ionization detector (GC-FID). GC column with 4.0mm~6.0mm radius and 2.0m~6.0m length is made out of copper or stainless. GC column has packing materials for measuring gases in exhaust gas; Porapak-QS is used for analysing carbon dioxide, and MS-5A is used for analysing nitrogen, oxygen and carbon monoxide.

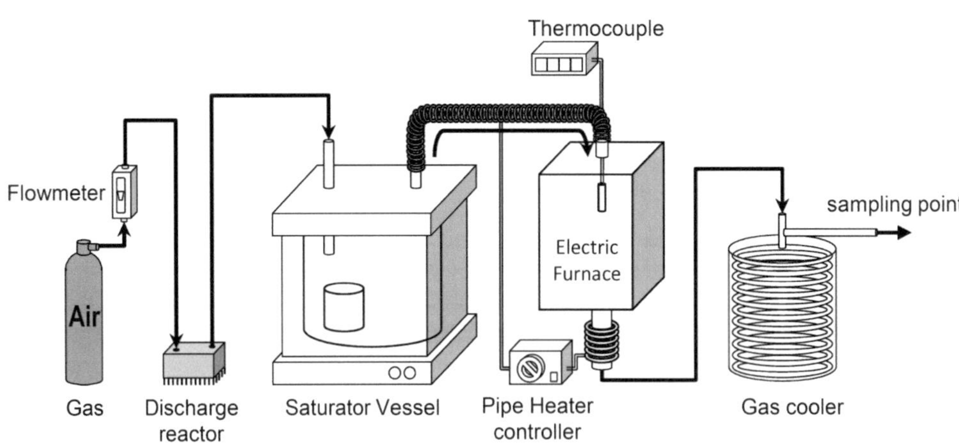

Figure 10. A schematic diagram of flow reactor system.

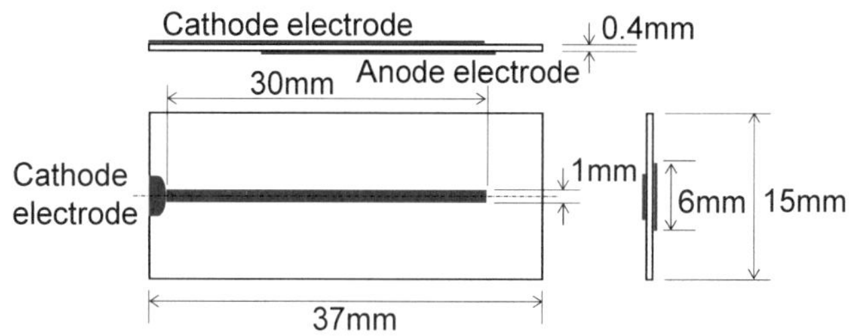

Figure 11. A schematic diagram of a surface discharge chip in discharge reactor.

5.2. Inversion rate of n-octane

Dry air or ozonized dry air is mixed to n-octane in saturator vessel and the gas mixture is combusted in the electric furnace varied from 200 to 800 degrees C for an hour. Combustion product gases are trapped and then analysed by GC-FID with diethyl ether as internal standard substance. Inversion rate of n-octane in vapour-phase combustion reactions is evaluated from n-octane concentration in exhaust gases. Concentration of n-octane is estimated by calibration curve, which determine the ratio of peak area and mass ratio of n-octane to that of the internal standard (diethyl ether) obtained from gas chromatogram. Furthermore, concentration of other main combustion product gases, such as nitrogen, oxygen, carbon dioxide and carbon monoxide, are analysed by GC-TCD as mentioned above and are appropriately quantified using calibration curve of peak area from gas chromatogram against known concentration.

Inversion rate of n-octane is derived from equation (1) that relates to the vapour pressure of n-octane (Antoine equation (2)) and the amount of n-octane (equation (3)).

$$I_r = 1 - \frac{C_o}{C_i} \tag{1}$$

I_r: inversion rate of n-octane [%]
C_o: concentration of unreacted n-octane in liquid reaction product [mol]
C_i: initial concentration of n-octane [mol]

$$LogP = A - \frac{B}{T+C} \tag{2}$$

P: vapour pressure [mmHg]
A: 6.924 (as Antoine constant of n-octane)
B: 11355.126 (as Antoine constant of n-octane)
C: 209.517 (as Antoine constant of n-octane)
T: temperature

$$F_{n\text{-}octane} = \frac{P_{n\text{-}octane} F_{air}}{P_{air}} \tag{3}$$

$F_{n\text{-}octane}$: flow rate of n-octane [ml/min]
$P_{n\text{-}octane}$: vapour pressure of n-octane [atm]
F_{air}: flow rate of air [ml/min]
P_{air}: vapour pressure of air [atm]

Inversion rate of n-octane in vapour-phase combustion against reaction temperature between 100 and 800 degree C is indicated in Figure 12. In comparison with dry air (without ozone), inversion rate tends to be higher when ozone is injected to saturator vessel. Especially, the rate indicates approximately 20% higher than without ozone around 300 degrees C of reaction temperature. Ozone gas is possible to react to n-octane in lower reaction temperature. Inversion rate of n-octane become similar in high temperature because it is suggested that half-life period of ozone tends to be shorter and then ozone becomes easily to decompose.

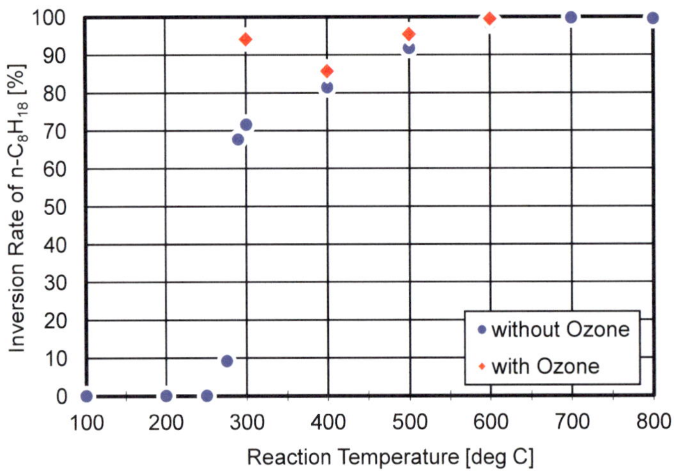

Figure 12. Relationship between reaction temperature and inversion rate of n-octane in vapour-phase combustion

5.3. Concentration of combustion product gases

On vapour-phase combustion reactions of n-octane with/without ozone, concentration of carbon dioxide and carbon monoxide is measured. Relationship between product gas concentration and reaction temperature is indicated in Figure 13. When with/without ozone flow into saturation vessel, concentration of carbon monoxide that is produced at the time of incomplete combustion gradually increases until 600 degrees C. Then concentration of carbon monoxide rapidly decreases in 700 degrees C of reaction temperature, and simultaneously, carbon dioxide increases very quickly in the case of combustion with dry air. Furthermore, in combustion with ozone-rich air, concentration of carbon dioxide indicates rapidly higher in 600 degrees C. It is supposed that n-octane with ozone combustion almost achieve complete combustion in 100 degrees lower than without ozone combustion. As a result, it is suggested that combustion reactions with/without ozone indicate a same reactions and proceed simultaneously with different reactions below.

Combustion reactions of n-octane without ozone

$$n-C_8H_{18}+O_2\rightarrow$$
$$intermediate(I)+H_2O+CO\rightarrow intermiediate\ (I)+H_2O+CO_2 \quad [300\sim500\ degrees\ C]$$

Combustion reactions of n-octane with ozone

$$n-C_8H_{18}+O_2\rightarrow$$
$$intermediate(I)+H_2O+CO\rightarrow intermiediate\ (I)+H_2O+CO_2 \quad [300\sim500\ degrees\ C]$$

Combustion reactions of n-octane with ozone

$$n-C_8H_{18}+O_3\rightarrow intermediate(II)+H_2O+CO_2 \quad [300\sim500\ degrees\ C]$$

Intermediate (I) and (II) in chemical equation indicate reaction products in combustion with/without ozone respectively. Combustibility of intermediate (II) is possibly higher than intermediate (I) and carbon dioxide directly generate by complete combustion. Therefore, concentration of carbon monoxide against reaction temperature indicates a similar tendency regardless of ozone injection. Furthermore, it is found that the intermediate product is obtained by GC-FID when mixture gas of n-octane and ozone or dry air is combusted. In addition, the intermediate product is NOT detected under the condition of only mixing vapour n-octane and ozone or dry air.

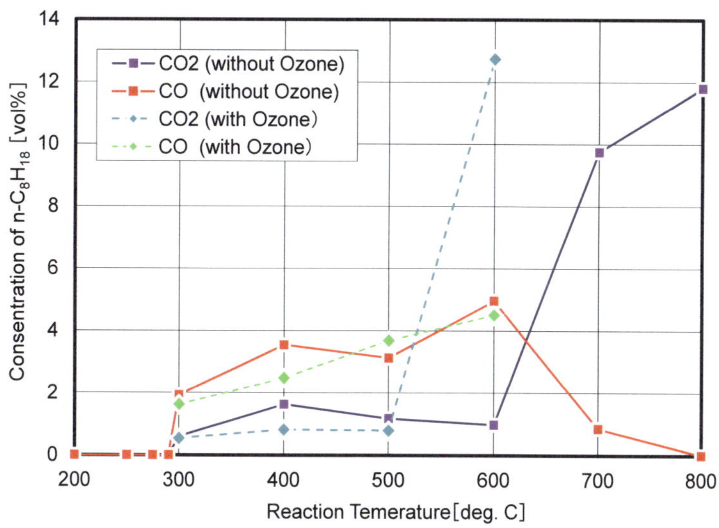

Figure 13. Relationship between product gas concentration and reaction temperature

5.4. Combustion heat

On vapour-phase combustion reactions of n-octane with/without ozone, maximum combustion heat in electric furnace is measured when reaction temperature is from 300 to

600 degrees C (Figure 14). Combustion heat with ozone indicates around 10 degrees C lower on each reaction temperatures compared with dry air (without ozone). One of the reasons of low combustion heat is possibly intermediate products caused by ozone.

Besides, the time to even out the combustion heat of electric furnace takes approximately 10 minutes in the case of ozone injection, although combustion heat without ozone takes more than 30 minutes. Ozone injection may contribute stable combustion because the time to even out the temperature of electric furnace also becomes shorter than without ozone.

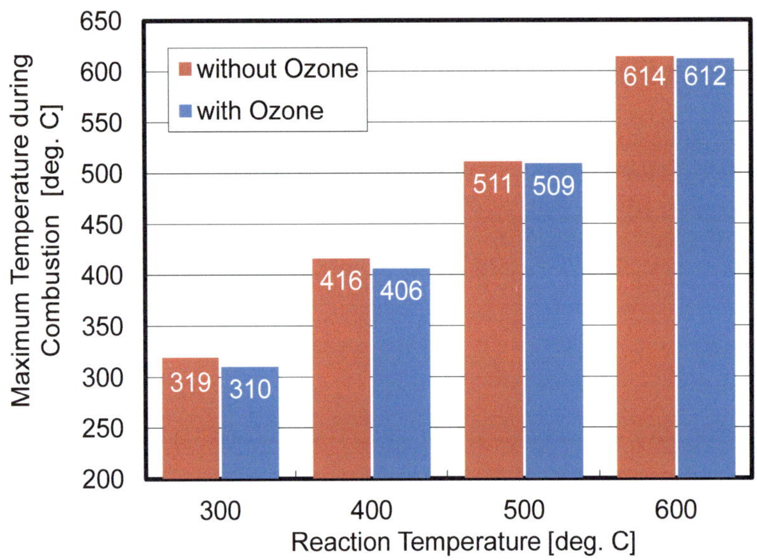

Figure 14. Relationship between maximum combustion heat and reaction temperature

6. Conclusion

FTIR spectra of vaporised hydrocarbon compounds (commercially available petrol and light diesel oil) mixed with discharged dry air were analysed for investigating intake stroke. Reaction products were found after ozone injection at around 1725cm^{-1} of FTIR spectra, and generation rate of it was strongly related to ozone concentration. Furthermore, it was found that thin ozone (0.76g/m^3) contributed to the generation of the reaction products, and ozone unused for the reaction was detected as extra at around 1058cm^{-1} and 2120cm^{-1}. Hydrocarbon combined with oxygen atom or molecule was detected by GC-MS and many of these reaction products had the FTIR spectra near 1725cm^{-1}. Extra ozone logically decomposed in the cylinder under the condition of the high pressure and high temperature. It was supposed that hydrocarbon was progressively ozonised in the cylinder, and these reaction products contributed to an efficient combustion.

It was found that the inversion rate indicates approximately 20% higher than without ozone especially in low temperature around 300 degrees C of reaction temperature. Furthermore, it is supposed that n-octane with ozone combustion almost achieve complete combustion in 100 degrees lower in comparison with air combustion. One of the reasons of high combustibility is possibly intermediate product obtained by combustion of mixture gas of n-octane and ozone.

Author details

Nobuya Hayashi
Kyushu University, Faculty of Engineering Sciences, Dep. of Engineering Science, Japan

Yoshihito Yagyu, Hideo Nagata, Hiroharu Kawasaki and Yoshiaki Suda
Sasebo National College of Technology, Japan

Seiji Baba
Densoken Co., Ltd., Japan

Acknowledgement

The authors would like to express our great thanks to Dr. ICHINOSE Hiromichi (Saga ceramics research laboratory). We also would like to thank the following people for their contribution to this research: Mr. YOSHIMURA Toyoji, Mr. TSUBOTA Takuya, Mr. YAMAGUCHI Masanori and Mr. IWAMOTO Genki. This research was partially supported by Grant in-Aid for Innovative Collaborative Research Projects of Sasebo National College of Technology, and the Education & Research Grant Program for Cooperation between Toyohashi University of Technology and National College of Technology (Kosen).

7. References

B. Kegl "Influence of biodiesel on engine combustion and emission characteristics", Applied Energy, 88(5), pp. 1803-1812 (2011)

Imagineering Inc.: JP 19-113570(A)

M. Wilk & A. Magdziarz, "Ozone Effects on the Emissions of Pollutants Coming from Natural Gas Combustion", Polish J. of Environ. Stud. 19(6), pp. 1331-1336 (2010)

M.Y. Kim, S.H. Yoon, B.W. Ryu & C.S. Lee, "Combustion and emission characteristics of DME as an alternative fuel for compression ignition engines with a high pressure injection system", Fuel, 87(12), pp. 2779-2786 (2008)

Nissan Motor Co., Ltd.: JP 15-113570(A)

NIST Standard Reference Database Number 69. (online), available from <http://webbook.nist.gov>

S. Verhelst & T. Wallner, "Hydrogen-fueled internal combustion engines", Progress in Energy and Combustion Science, 35(6), pp. 490-527 (2009)

Sun chemical Co., Ltd.: JP 7-301160

T. Tachibana, K. Hirata, H. Nishida, & H. Osada, "Effect of ozone on combustion of compression ignition engines", Combustion Flame, vol. 85(3), pp.515–519 (1991)

T.M.I. Mahlia, S. Tohno, T. Tezuka, "A review on fuel economy test procedure for automobiles: Implementation possibilities in Malaysia and lessons for other countries", Renewable and Sustainable Energy Reviews, 16(6), pp. 4029-4046 (2012)

Chapter 3

Syngas Application to Spark Ignition Engine Working Simulations by Use of Rapid Compression Machine

Eliseu Monteiro, Marc Bellenoue, Julien Sottton and Abel Rouboa

Additional information is available at the end of the chapter

http://dx.doi.org/10.5772/48088

1. Introduction

It has become more and more urgent to find alternative fuels or alternative sources of energy for transportation systems as reserves of standard fossil fuels are decreasing very rapidly. Therefore it is necessary to find alternative fuels to be used in the standard internal combustion engine to bridge this gap. Biomass is considered as the renewable energy source with the highest potential to contribute to the energy needs of modern society for both the developed and developing economies world-wide [1]. Energy from biomass, based on short rotation forestry and other energy crops, can contribute significantly towards the objectives of the Kyoto Agreement in reducing the greenhouse gases emissions and to the problems related to climate change [2]. The gasification of biomass allows the production of a synthesis gas or "syngas," consisting primarily of H_2, CO, CH_4, CO_2 and N_2 [3]. The specific composition depends upon the fuel source and the processing technique. These substantial variations in composition and heating value are among the largest barriers toward their usage.

The main advantage that comes from the use of syngas in SI engines over the conventional liquid, petroleum-based fuels is the potential for increased thermal efficiency [4]. This is attributed to the relatively high compression ratios permitted, usually by converting Diesel engines for gaseous fuel operation in the SI mode [5], since CO and CH_4 are characterized by high anti-knock behavior [6]. On the contrary, the relatively increased end-gas temperature, which the fast flame propagation rate of H_2 can produce during combustion and can be responsible for knock onset, is compensated for by the presence of diluents in the fuel (N_2 and CO_2). Their effect on combustion is to lower flame speed and so decrease the in-cylinder pressures and temperatures. The moderation of peak gas temperatures during combustion, attributed to this feature, has also a reduction effect on NO_x emissions [8]. Besides, the drawback of reduced power output using fuels with relatively low heating values can be partially balanced by turbo-charging the engine. Towards the direction of minimizing this power derating when, for example, syngas with low heating value equal to 4–6 MJ/Nm^3 is used instead of natural gas with low heating value of approximately 30 MJ/Nm^3, contributes the fact that the syngas stoichiometric air–fuel ratio is about 1.2 compared with the value of 17 for the natural gas case. Thus, the energy content per unit quantity of mixture (air + fuel) inducted to the cylinder is only marginally lower when using syngas, compared with the corresponding natural gas case [9].

In single-cylinder or multi-cylinder engines, it is very difficult to control the combustion, because parameters are coupled with each other under engine operating conditions. For this reason, the use of a rapid compression machine (RCM) allows to elucidate about combustion characteristics and visualize combustion phenomena [10-15].

In this chapter, two typical mixtures of H_2, CO, CH_4, CO_2 and N_2 have been considered as representative of the producer gas coming from wood gasification, and its turbulent combustion at engine-like conditions is made in a rapid compression machine designed to simulate the thermodynamic cycle of an engine, particularly compression and expansion strokes, in order to improve current knowledge and provide reference data for modeling and simulation with the objective of its application in stationary energy production systems based on internal combustion engines.

2. Materials and methods

2.1. Syngas

Gasification is the thermo-chemical conversion of a carbonaceous fuel at high temperatures, involving partial oxidation of the fuel elements. The result of the gasification is a fuel gas - the so-called syngas - consisting mainly of carbon monoxide (CO), hydrogen (H_2), carbon dioxide (CO_2), water vapor (H_2O), methane (CH_4), nitrogen (N_2), some hydrocarbons in very low quantity and contaminants, such as carbon particles, tar and ash.

Syngas-air mixtures and methane-air mixture are prepared in bottles by means of partial pressure method, and then the mixture is prepared within the chamber by adding syngas and air at specified partial pressures. The purity of the gases is in all cases at least 99.9%. The typical syngas compositions are shown in the Table 1.

Gasifier	H_2	CO	CO_2	CH_4	N_2
Updraft	11	24	9	3	53
Downdraft	17	21	13	1	48

Table 1. Syngas compositions (% by volume)

The simplified chemical reaction that expresses the stoichiometric combustion of syngas for syngas typical compositions is [16]:

$$aH_2 + bCO + cCH_4 + dCO_2 + eN_2 + \left(\frac{a}{2} + \frac{b}{2} + 2c\right)(O_2 + 3.76N_2) \rightarrow \\ (b+c+d)CO_2 + (a+2c)H_2O + \left[e + 3.76\left(\frac{a}{2} + \frac{b}{2} + 2c\right)\right]N_2 \quad (1)$$

Where *a,b,c,d,* and *e* are molar coefficients. Five experiments for each mixture and ignition timing were performed in order to assure the good reproducibility of the signals. For single compression the maximum difference between peak pressures is: 1.8 bar for ignition at top dead center (TDC) (70 bar on average, which represents an error of 2.5%); 0.8 bar for ignition timing at 5.0 ms before TDC (BTDC) (69.2 bar on average representing an error of 1.1%); 0.5 bar for ignition timing at 7.5 ms BTDC (68.8 bar on average, which represents an error of 0.7%); 0.5 bar for ignition timing at 12.5 ms BTDC (68.4 bar on average representing an error of 0.7%). For compression-expansion the maximum difference between peak pressures is: 0.2 bar for ignition timing at 5 ms BTDC (30 bar on average representing an error of 0.7%); 0.1 bar for ignition timing at 7.5 ms BTDC (38 bar on average, which represents an error of 0.03%); 0.8 bar for ignition timing at 12.5 ms BTDC (47.5 bar on average representing an error of 1.7%).

2.1.1. *Syngas flammability limits*

The flammability limit is the most widely used index for representing the flammability characteristics of gases. In accordance with generally accepted usage, the flammability limits are known as those regions of fuel–air ratio within which flame propagation can be possible and beyond which flame cannot propagate. And there are two distinct separate flammability limits for the fuel–air mixture, namely, the leanest fuel-limit up to which the flame can propagate is termed as lower flammability limit (LFL), and the richest limit is called as upper flammability limit (UFL).

The purpose of this study is to determine the flammability limits of syngas-air mixtures as guidance for stationary energy production systems. As the conditions in an energy production scenario are different, namely in terms of pressure, some apparatus characteristics referred in the standards described above are not followed. The flammable region obtained will be narrower than the actual flammable region.

In this context, prior to our experimental work, various measurements of the flammable region of syngas-air mixtures with a specific ignition system providing ignition energy of 45 mJ and initial conditions of 293 K for temperature and 1.0 bar and 3.0 bar for pressure.

The syngas-air mixtures were prepared within the spherical chamber using two 10L bottles previously prepared with the syngas composition and another with compressed atmospheric air by the partial pressure method. The equivalence ratio was varied in 0.1 steps. The flammability limits using for different initial conditions of pressure is shown in figures 1-2. In the boundaries of the flammable region ten shots were made in order to get the ignition success.

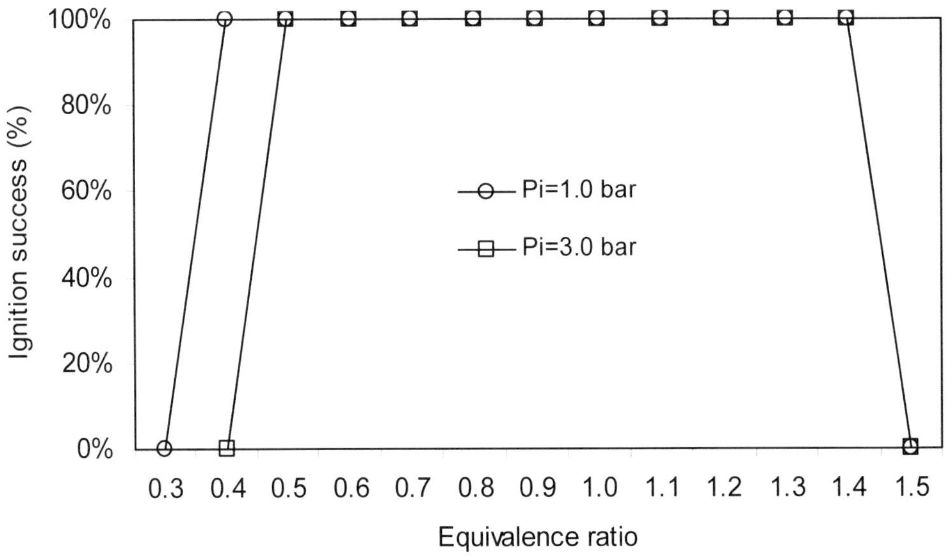

Figure 1. Flammability limits of the updraft syngas-air mixture.

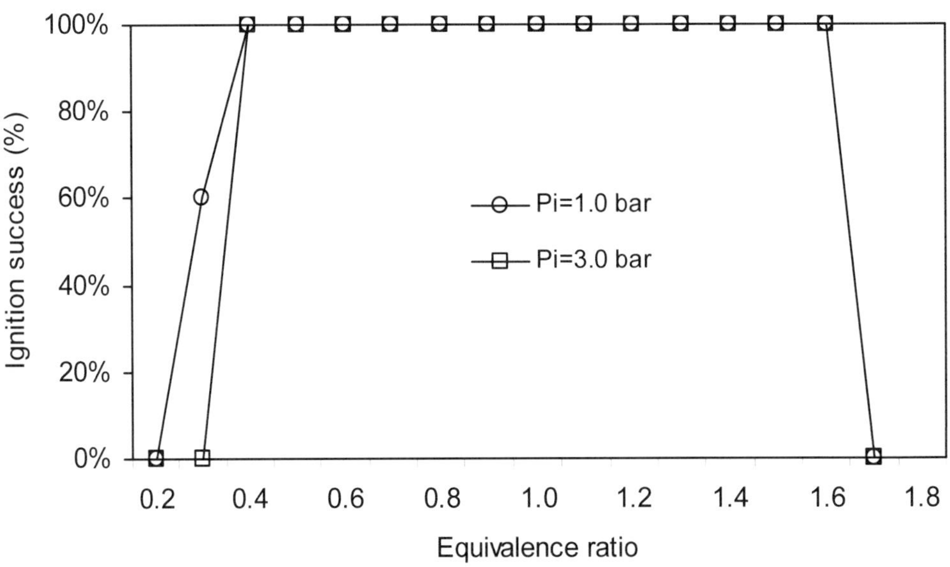

Figure 2. Flammability limits of the downdraft syngas-air mixture.

Figure 1 shows that the flammable region for the updraft syngas-air mixture for 1.0 bar and 293K is between 0.4 and 1.4 equivalence ratios. Increasing the initial pressure to 3.0 bar, the lean flammability limit is reduced to ϕ=0.5.

Figure 2 shows that the flammable region for the downdraft syngas-air mixture for 1.0 bar and 293K is between 0.3 and 1.6 equivalence ratios. Pressure increase for ϕ=0.3 is only 20% of the initial pressure. Increasing the initial pressure to 3.0 bar, the lean flammability limit is reduced to ϕ=0.4.

2.1.2. Burning velocity

The stretched burning velocity, S_u, of the propagating flame is calculated by the following expression [17]:

$$S_u = \frac{r_v}{3(P_v - P_i)} \left(\frac{P_i}{P}\right)^{\frac{1}{\gamma}} \left(1 - \left(\frac{P_i}{P}\right)^{\frac{1}{\gamma}} \frac{P_v - P}{P_v - P_i}\right)^{-\frac{2}{3}} \frac{dP}{dt} \tag{2}$$

Where P_i and P_v are the initial and maximum pressure, respectively. γ is the specific heat ratio of the mixture and r_v is the radius of the combustion chamber.

The simultaneous change in the pressure and temperature of the unburned mixture during a closed vessel explosion makes it necessary to rely on correlations which take these effects into account like the one proposed by [18]:

$$S_u = S_{u_0} \left(\frac{T}{T_0}\right)^{\alpha} \left(\frac{P}{P_0}\right)^{\beta} \tag{3}$$

Where, T_0 and P_0, are the reference temperature and pressure, respectively. α and β, are temperature and pressure exponents, respectively and S_{u0} the reference burning velocity.

The correlations of laminar burning velocities of the typical syngas compositions of the type $S_u = S_{u0}(T/T_0)^{\alpha}(P/P_0)^{\beta}$ under stoichiometric conditions are [16]:

$$S_u = 0.303(T/T_0)^{1.507}(P/P_0)^{-0.259} \tag{4}$$

$$S_u = 0.345(T/T_0)^{1.559}(P/P_0)^{-0.156} \tag{5}$$

where, T is the temperature, P the pressure and the index 0 represents reference conditions (1.0 bar, 293 K).

2.2. Rapid compression machine

The experiments were conducted in a rapid compression machine (RCM) at the Institute P' of the ENSMA. This RCM has been designed in such a manner that the piston velocity evolves similarly as it does in a real engine. Volumetric compression ratio of the RCM can be varied from ε=9.1 to ε=18.8 by changing the clearance volume. The RCM features a square cross/section with rounded corners piston (50×50, r=3.6 mm), allowing flat windows to be mounted on lateral sides of the chamber; this enables direct visualizations and planar laser sheet measurements within the whole dead volume (Fig. 3).

The piston is equipped with a squared sealing ring but also with a square shaped guiding ring, to avoid asymmetric formation of corner vortex. Moreover the RCM features a long compression stroke (S=419 mm), providing a wider visualization window at top dead center (TDC) for a given compression ratio. Compression of reactive mixtures is obtained as follows. A hydraulic cylinder sets a cam into motion. The horizontal translation is transformed into a vertical motion via a guiding wheel. The RCM is equipped with a return cylinder to keep the contact between the cam and the guiding wheel; this ensures that the volume of the chamber is well stabilized at TDC.

The RCM is fitted with a heating system in the chamber walls to vary the initial temperature at bottom dead center (BDC) between 293 and 373 ±1.5K.

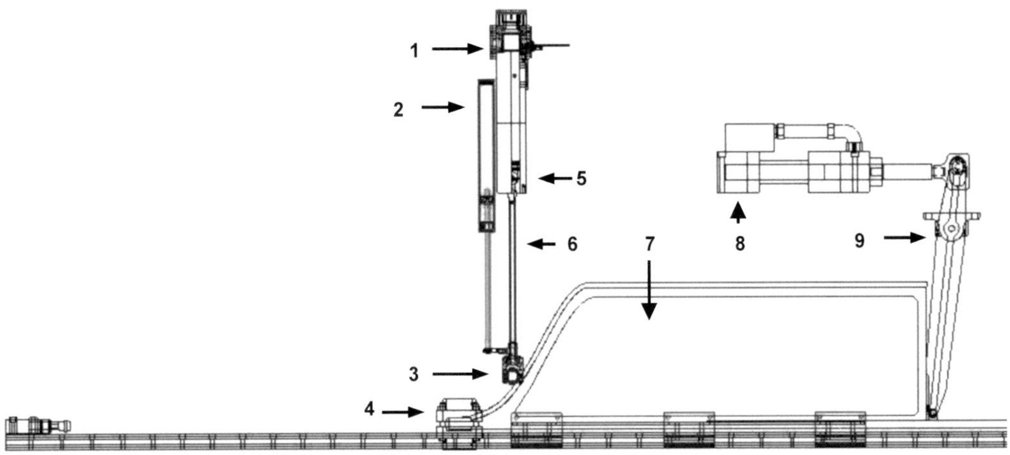

Figure 3. RCM scheme. (1) Cylinder/combustion chamber, (2) return cylinder, (3) guiding wheel, (4) Brake system, (5) Piston, (6) Connecting rod, (7) Cam, (8) hydraulic cylinder, (9) lever.

The mechanical part of the RCM is based on the principle of the catapult. A cylinder, commanded by a high-speed servo valve (response time of approximately 5 ms) and supplied with a hydro-electric power station allowing a flow of 400 l/min, drives a steel rod of large dimensions (length 775 mm, thickness 50 mm, H section) which in turns drive a carriage of 40 kg at high speed installed horizontally on rails. This carriage is equipped with a cam whose profile actuates a roller that supports the piston vertical movement. The profile is calculated to reproduce the movement of an engine running at around 700 rpm. It was necessary to install hydraulic brakes (ordered in a mechanical way, by safety) on the way of the carriage in order to stop it, which passes from 0 to 40 km/h and then from 40 km/h to 0. Everything will take 60 ms, which explains the 5 meters length of the machine.

The RCM is equipped with various measurement means: a laser sensor to measure displacement, inductive sensors positioned along the axis of the piston to start the optical instrumentation and to generate the spark, a sensor to measure the dynamic pressure in the combustion chamber, as well as a thermocouple used during special tests where the chamber is heated, and controlled in temperature. The chamber is equipped with valves intended for the draining and filling of gas mixtures, as well as a secondary cylinder to create controlled aerodynamic effects representative of those found in engines (swirl movement, tumble or homogeneous turbulence).

The RCM control is managed by a PC. The measuring signs (pressure, piston position, wall temperature, heat flux, etc.) are registered simultaneously by a data acquisition system (National Instrument 6259) and integrated in the interface. Also some RCM controlling parameters (brakes pressure, hydraulic pressure, piston position, contact cam/lever, etc.) are taking into account for security reasons. The interface also controls the signals of the lasers and camera.

In an ideal spark-ignited internal combustion engine one can distingue three stages: compression, combustion and expansion. The entire pressure rise during combustion takes place at constant volume at TDC. However in an actual engine this does not happen as well as in the RCM. The pressure variation due to combustion in a compression and expansion rapid compression machine is shown in figure 4 where three stages of combustion can be distinguished.

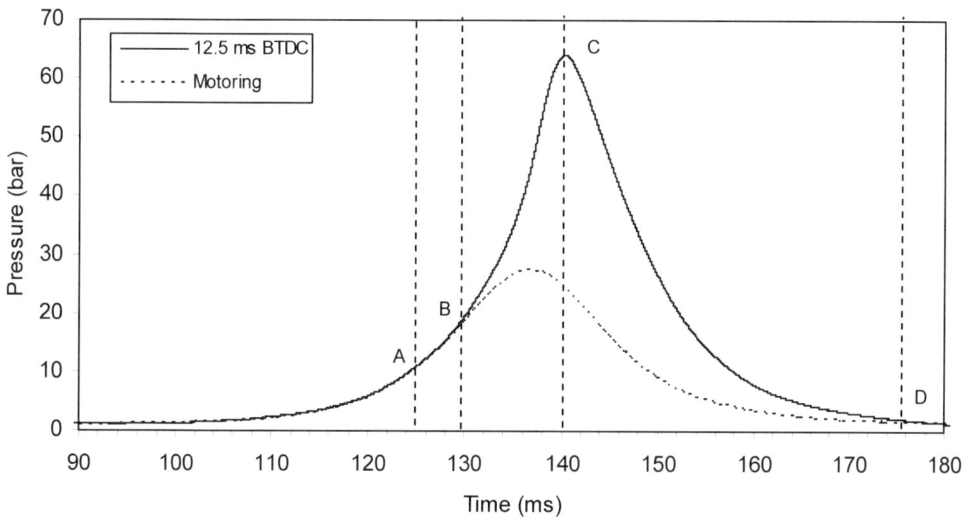

Figure 4. Stages of combustion in a RCM.

In this figure, A is the point of passage of spark, B is point at which the beginning of pressure rise can be detected and C the attainment of peak pressure. Thus, AB represents the first stage, BC the second stage and CD the third stage [19].

The first stage is referred to as ignition lag or preparation phase in which growth and development of a self-propagating nucleus of flame takes place. This is a chemical process depending upon both pressure and temperature and the nature of the fuel. Further, it is also dependent of the relationship between the temperature and the rate of reaction.

The second stage is a physical one and is concerned with the spread of the flame throughout the combustion chamber. The starting point of the second stage is where the first measurable rise of pressure is seen, i.e. the point where the line of combustion departs from the compression line (point B). This can be seen from the deviation from the compression (motoring) curve.

During the second stage the flame propagates practically at a constant velocity. Heat transfer to the cylinder wall is low, because only a small part of the burning mixture comes in contact with the cylinder wall during this period. The rate of heat release depends largely of the turbulence intensity and also of the reaction rate which is dependent on the mixture composition. The rate of pressure rise is proportional to the rate of heat release because during this stage, the combustion chamber volume remains practically constant (since the piston is near the TDC).

The starting point of the third stage is usually taken at the instant at which the maximum pressure is reached (point C). The flame velocity decreases during this stage. The rate of combustion becomes low due to lower flame velocity and reduced flame front surface. Since the expansion stroke starts before this stage of combustion, with the piston moving away from the TDC, there can be no pressure rise during this stage.

2.2.1. Aerodynamics inside the RCM

Although in principle RCM simulates a single compression event, complex aerodynamic features can affect the state of the reacting core in the reaction chamber. Previous studies [20, 21] have shown that the motion of the piston creates a roll-up vortex, which results in mixing of the cold gas pockets from the boundary layer with the hot gases in the core region. However, substantial discrepancies have been observed between data taken from different rapid compression machines even under similar conditions of temperature and pressure [22]. These discrepancies are attributed partly to the different heat loss characteristics after the end of the compression stroke and partly to the difference in aerodynamics between various machines. The effect of aerodynamics is particularly more complicated because it does not show up in the pressure trace and it may lead to significant temperature gradients and ultimately to the failure of the adiabatic core hypothesis.

The aerodynamics inside a rapid compression machine is highly unsteady in nature; it plays a role in pre-ignition through turbulent mixing, but also because it drives the evolution of the temperature distribution. The characterization of the temporal evolution of the flow, and quantify the distribution and turbulence intensity is made using an inert gas N_2 to simplify the diagnosis and avoiding the disruption of PIV images by possible oxidation of unwanted particles. The flow remains representative of the reactive case when the heat release is negligible. Measurements on the total extent of the clearance volume and at the center of the chamber were made.

2.2.1.1. Velocity fluctuations

The study of turbulent flows is generally based on the Reynolds decomposition, where the instantaneous velocity (U) is decomposed into an averaged (<U>) part and a fluctuating (u) part: $U = \langle U \rangle + u$.

In most cases, a global average is used to estimate mean velocity. Using this approach in an engine results in substantial overestimation of the turbulent intensity that can reach a factor of 2 [23]. Indeed, the cyclical fluctuations of the overall movement (such as large eddy scale movement) are included in the fluctuating field as well as fluctuations in velocity caused by the turbulent nature of the flow.

Instantaneous. velocity

Figure 5 shows the time evolution of the velocity field during an inert gas compression. It is observed 10 ms BTDC a laminar one-dimensional compression flow. A zone of high velocities (5 to 8 m/s), where the flow is turbulent, that come in the center of the clearance volume 5 ms after. The laminar flow of this zone becomes two-dimensional and diverging to the walls. The turbulent zone reaches TDC and occupies a large part of the chamber at that moment. The flow in this zone is structured by two counter-rotating vortices, which is consistent with the literature where the movement of the piston brings the gas from the side wall toward the center of the chamber, forming vortices on the corners. These vortices then move to the side walls and after down the chamber. Simultaneously, the maximum velocity of the flow gradually decreases, and the size of the 'laminar' zone observed at the end of compression decreases. The disappearance of this zone occurs approximately 17 ms after TDC, although some low velocity zones remain. 40 ms after TDC, the corner vortices are replaced by a fragmented and highly three-dimensional flow.

The coexistence of laminar and turbulent regions is characteristic of MCR flat piston flow, where the gases are at rest before compression. One can observe a certain asymmetry in vortices velocity, with lower values at TDC and close to the walls (figure 5). This asymmetry reflects the exchange of kinetic energy that occurs in the strain layer between the vortex and the zone of lower flow. The velocity gradient direction at the zone interface may also be parallel to the mean flow, as is in the case of few milliseconds before TDC. In this case, if the inertia of the high speed zone regrowth clearly the core zone, the turbulent nature of the

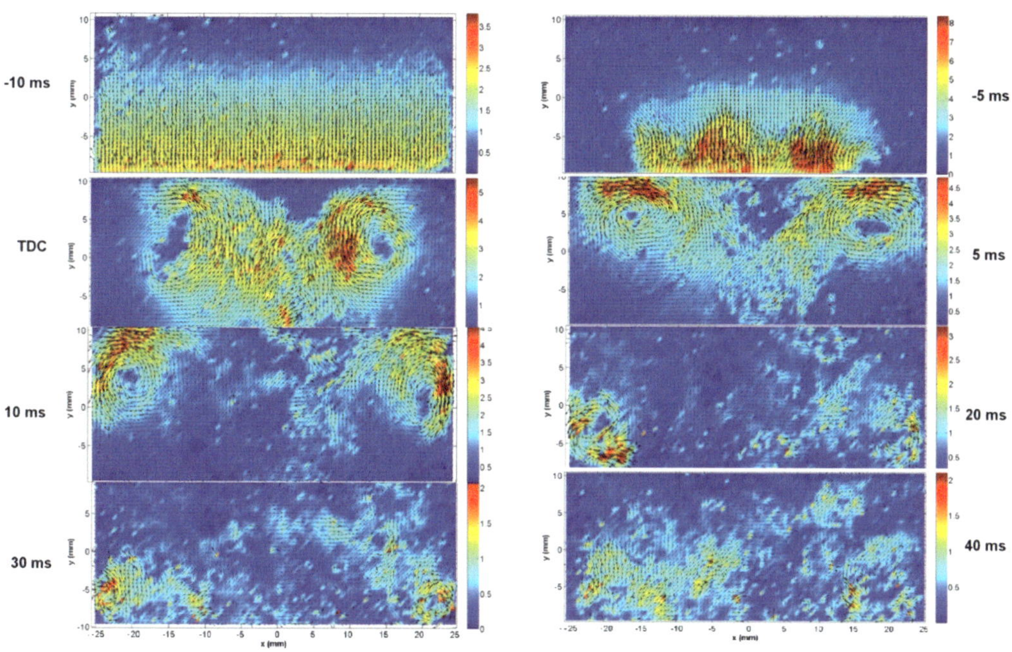

Figure 5. Inert compression velocity fields (m/s) in the RCM [15].

flow at the interface is also likely to accelerate the decrease in the extent of the core zone. This emphasis the existence of two ranges of scales associated with mixing phenomenon: those of the overall movement, and those of turbulence. Moreover, the overall velocity of the movement decreases rapidly if the flow stops. Thus, reflecting the kinetic energy transfer from large scales to the turbulent scales.

2.2.1.2. Analysis of the flow at the chamber core

The whole movement has been analyzed, and an initial assessment of turbulent fluctuations was provided from the whole filed measurements. Specific field measurements are now exposed to evaluate the properties of turbulence in detail. The turbulent characteristics are evaluated thanks to particle image velocimetry (PIV) measurements with time resolution of 5 kHz along one field of 13x13 mm and image resolution of 512x512 pixels. The investigated zone is close to the center of the chamber (1.5 mm to the left), where the mean and fluctuating velocities remain relatively high along a 10 ms period after TDC. Figure 6 shows the fluctuation velocity components in this zone. It is observed that both velocity components fluctuations decrease after TDC with similar amplitude. The maximum of flow velocity is obtained 2-3 ms before TDC with values around 0.65 m/s. It is followed by a rapid decrease to around 0.4 m/s that reflect the overall decrease of the convection of the fastest zones outside the measured field.

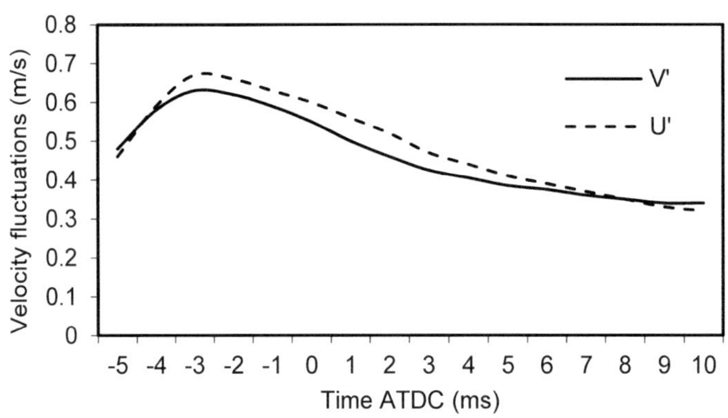

Figure 6. Time evolution of the variation of velocity fluctuations [15].

Both velocity components fluctuations decrease after TDC with similar amplitude. The kinetic energy (k) is evaluated only from two velocity components (u,v) as follows:

$$k = \sqrt{u'^2 + v'^2} \quad (6)$$

It is therefore, slightly underestimated (~20%) due to the lack of the third component. The turbulent intensity I obtained by:

$$I = \frac{u'}{U} \quad (7)$$

Where u′ is the root mean square (rms) of the turbulent velocity fluctuations and U the mean velocity (Reynolds averaged).

Figure 7 represents the kinetic turbulent energy. The maximum of kinetic energy is obtained 2-3 ms before TDC. It is followed by a rapid decrease that reflects both the overall decrease of the turbulent kinetic energy but also the convection of the fastest zones outside the measured field.

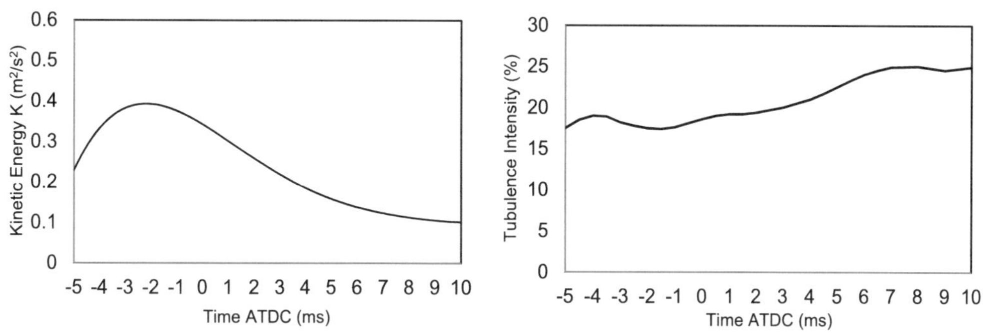

Figure 7. Kinetic energy (left) and turbulence intensity (right) [15].

The turbulent intensity is moderate, with a value of about 20% with minor variations over time. One should remind that this value corresponds to a high velocity turbulent zone.

3. Results and discussion

The RCM can work on two distinctive modes: single compression stroke and compression and expansion strokes.

Single compression is generally used for the study of high pressure auto-ignition of combustible mixtures as it gives direct measure of ignition delay [11]. When the interest is the heat transfer to the walls then it is usually used an inert gas, with equal adiabatic coefficient as the reacting mixture, as a test gas. In this work instead of an inert gas a stoichiometric syngas-air mixture was used out of auto-ignition conditions in order to provide data for the thermal model simulation.

Compression and expansion strokes simulate a single engine cycle of an internal combustion engine under easily controlled conditions and a cleaner environment than the traditional internal combustion engine.

3.1. Single compression

Figures 8-9 show RCM experimental pressure histories of stoichiometric syngas-air mixtures for various spark times and compression ratio ε=11. Four ignition timings were tested: TDC, and 5.0, 7.5 and 12.5 ms before TDC, respectively.

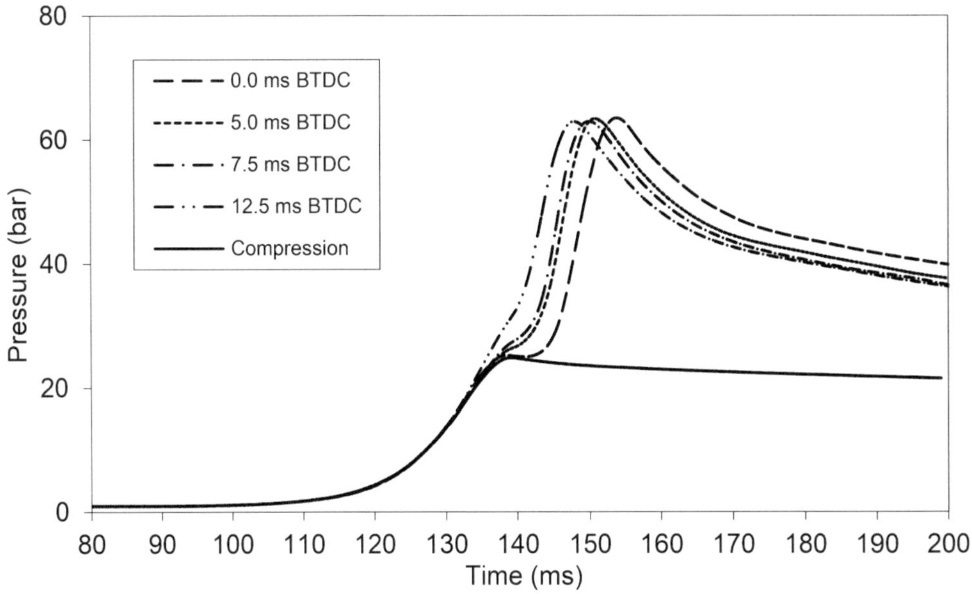

Figure 8. Pressure versus time for stoichiometric updraft syngas-air mixture at various spark times.

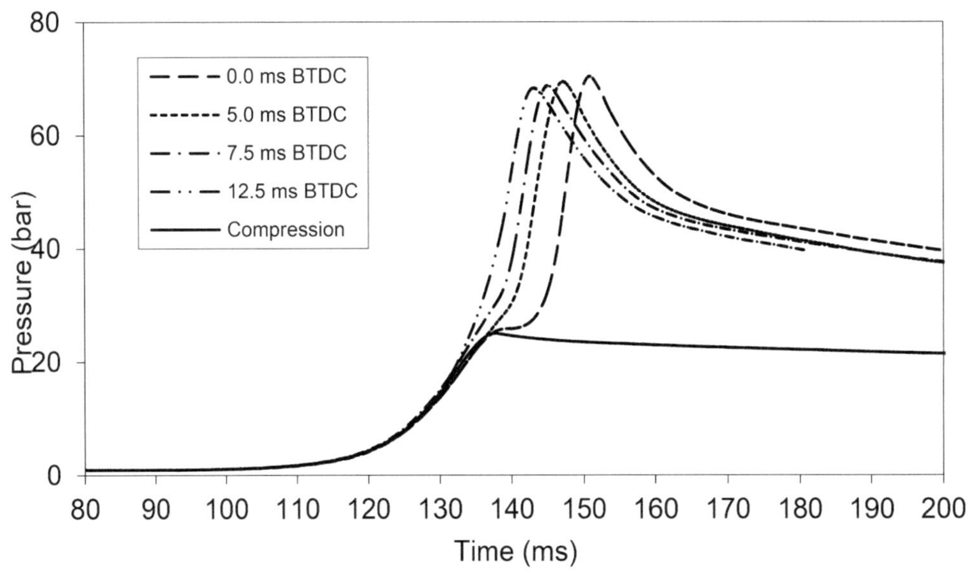

Figure 9. Pressure versus time for stoichiometric downdraft syngas-air mixture at various spark times.

In both syngas cases, peak pressure decreases as the ignition time increases. When ignition is made at TDC, combustion will occur at constant volume, in the clearance volume. In this case, one has higher initial pressure and temperature and no influence of the flow of the fresh mixture being compressed by the piston movement, which reduces turbulence and, in turns, the heat transfer. As far as ignition timing concerns, the deviation from TDC allows lower initial pressure and temperature for combustion in the compression stroke and consequently lower peak pressure. It is also observed a reduction in the pressure gradient after TDC, which means that the heat released by combustion of syngas-air mixtures is not enough to keep the same pressure gradient. One can observe that the pressure gradient is kept for ignition timing of 12.5 ms BTDC.

As mentioned above, experiments with stoichiometric methane-air mixture were also performed for comparison reasons and the results are shown in Fig. 10.

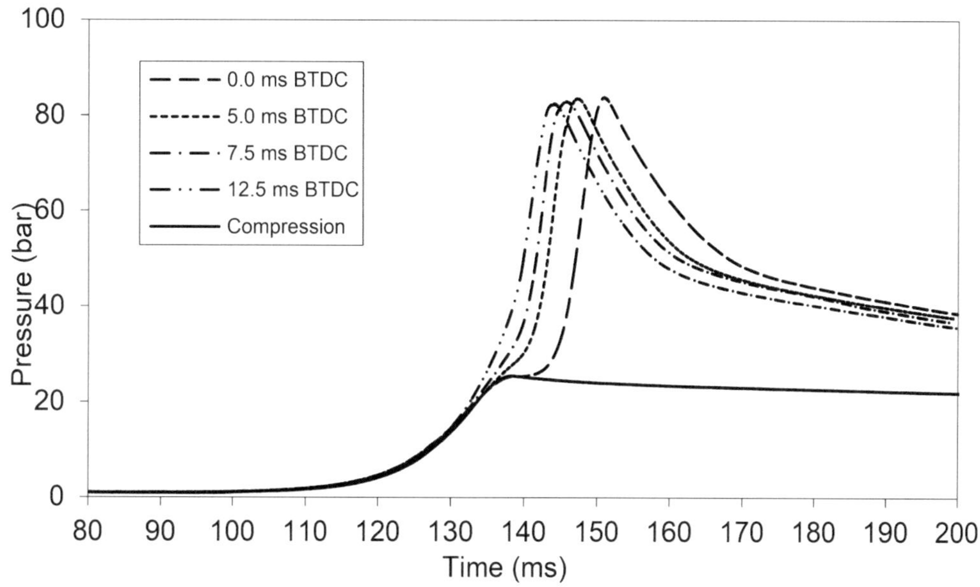

Figure 10. Pressure versus time for stoichiometric methane-air mixture at various spark times.

This mixture follows the same behaviour of the typical syngas compositions nevertheless with higher pressures. For these results contributes the fact that the syngas stoichiometric air–fuel ratio ranges between 1.0 (downdraft) and 1.12 (updraft) compared with the value of 9.52 for methane. Taking into account that the RCM chamber has 1.0 Liter, the energy introduced in to the chamber is 2.60 kJ in the updraft case, 2.85 kJ in the downdraft case and 3.38 kJ in the methane case for stoichiometric conditions. These values are in agreement with the obtained cylinder pressures, however not proportional in terms of peak pressures due to the influence of heat losses. These are mainly dependent of the quenching distance as well as thermal conductivity of the mixture. The higher burning velocity of methane (see Ref. [24]) compared to syngas compositions also cause a more intensified convection.

3.1.1. Direct light visualization from chemiluminescence emission

Direct visualizations of the flame propagation in a RCM are shown in Fig. 11 for updraft syngas and for ignition at TDC and 12.5 ms BTDC, to clearly establish differences.

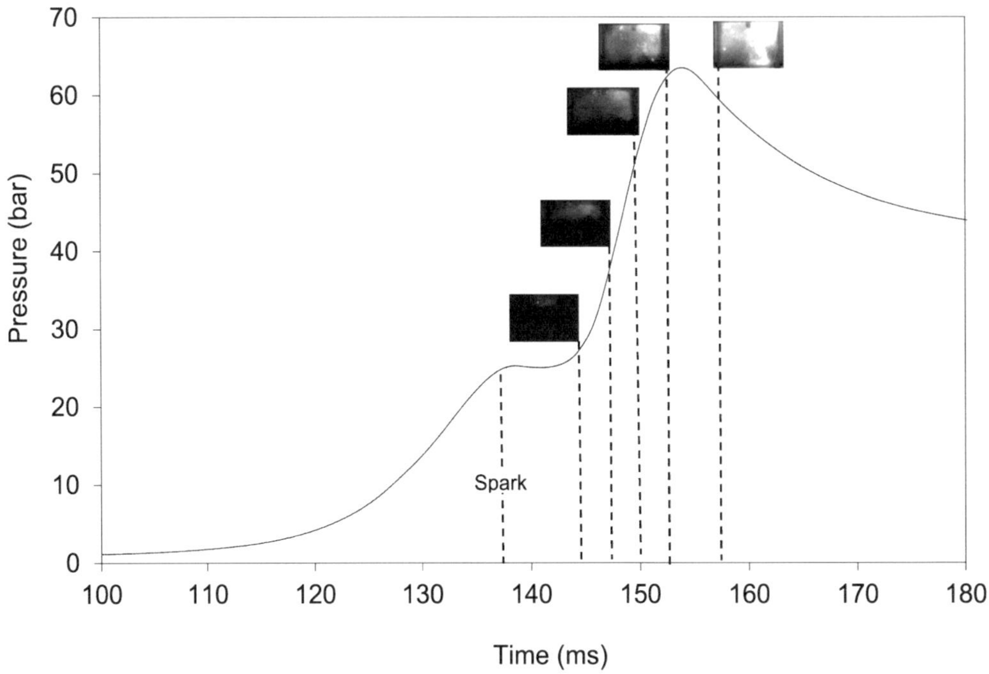

(a)

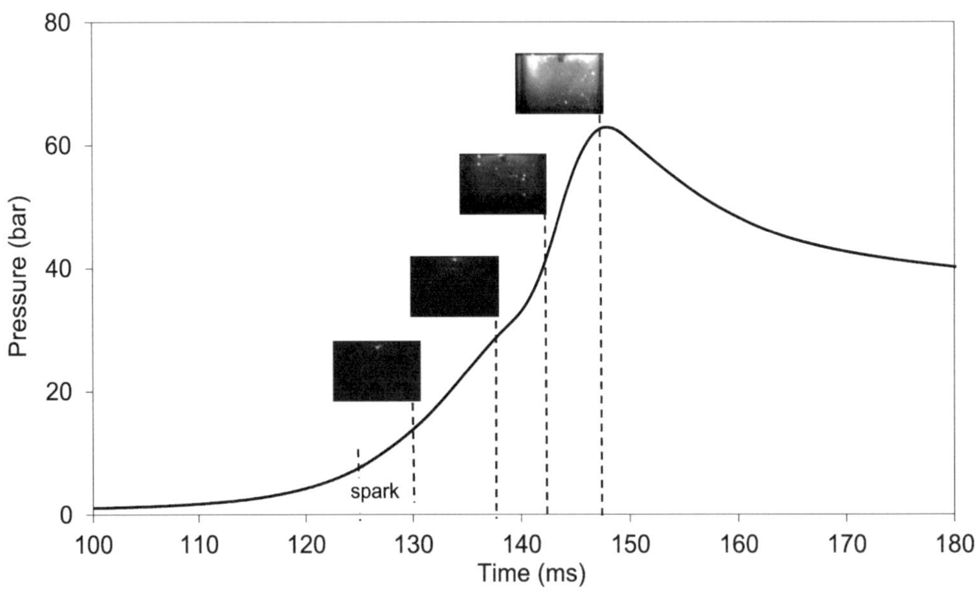

(b)

Figure 11. Direct visualization of stoichiometric updraft syngas-air mixtures combustion in a RCM. (a) Ignition at TDC; (b) Ignition at 12.5 ms BTDC [25].

When the ignition is made at TDC the combustion occurs at constant volume, in the clearance volume. Direct visualizations show an explosion with fast and turbulent flame propagation with combustion duration of about 17.5 ms. In opposite, the initial phase of combustion shows a quasi-spherical relatively smooth flame kernel specially for ignition timing of 12.5 ms BTDC. The flame kernel propagation is laminar and at some point experience flattening due to piston movement reaching the TDC. After that, a change to constant volume combustion occurs. The whole combustion duration is around 22.5 ms, which represents a remarkable increase compared with the full constant volume combustion. The deviation of the spark plug from TDC allows lower initial pressure and temperature for combustion in the compression stroke and lower turbulence intensity [12].

3.2. Compression –expansion

Figures 12-14 show RCM experimental pressure histories of stoichiometric syngas-air mixtures and methane-air for various spark times and compression ratio ε=11. The ignition timings tested were 5.0 ms, 7.5 ms and 12.5 ms BTDC.

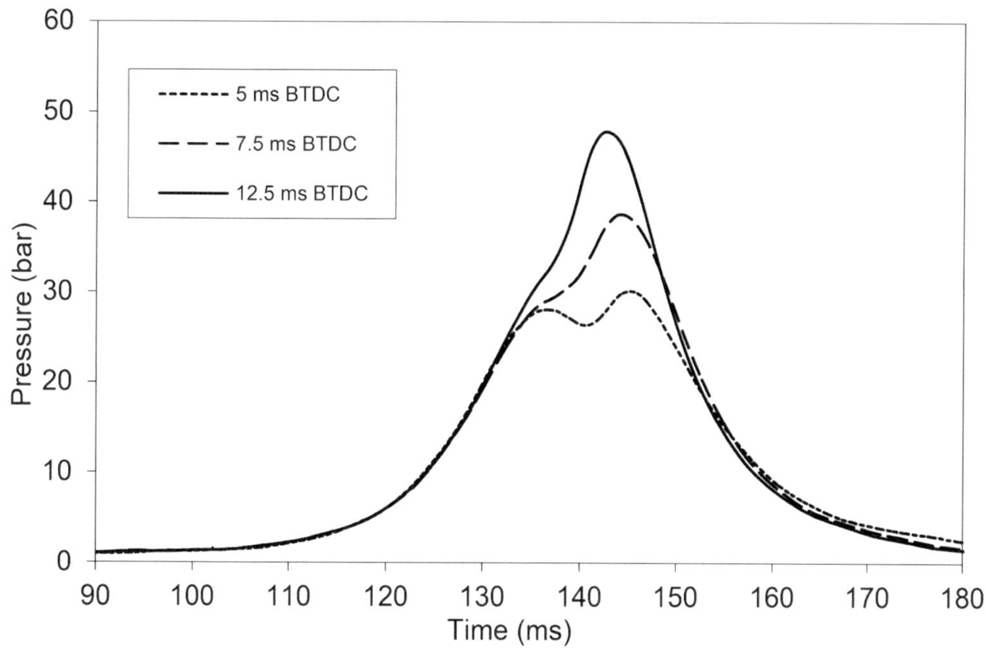

Figure 12. Pressure versus time for stoichiometric updraft syngas-air mixture at various spark times.

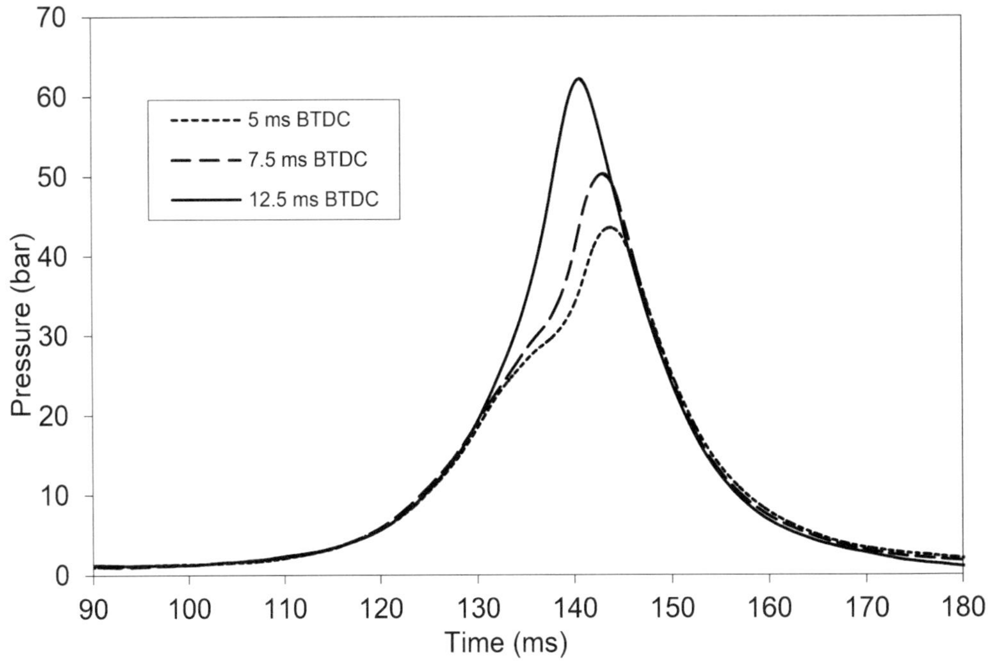

Figure 13. Pressure versus time for stoichiometric downdraft syngas-air mixture at various spark times.

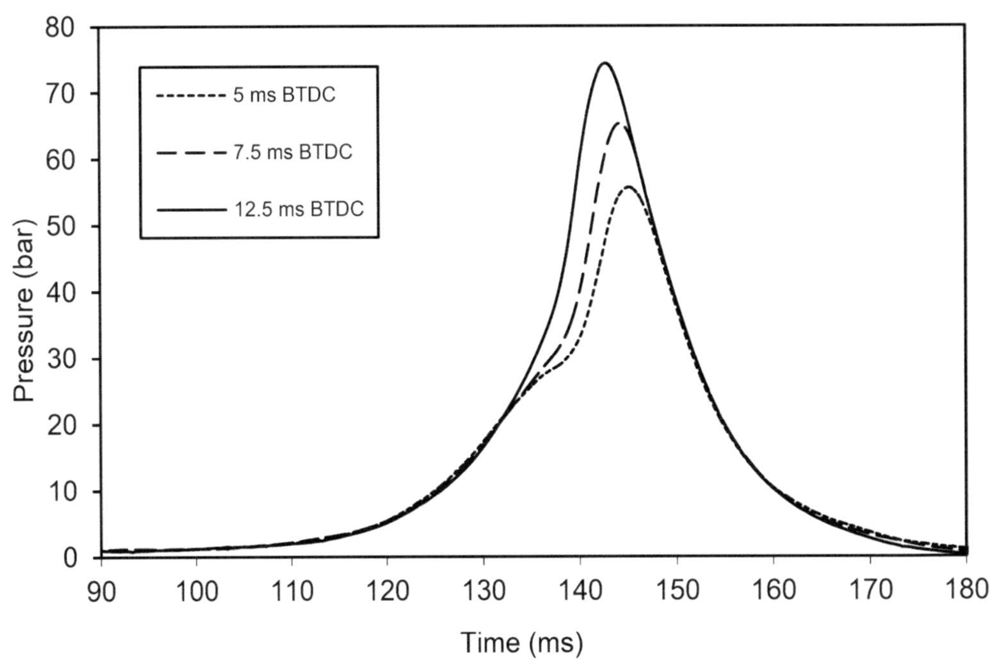

Figure 14. Pressure versus time for stoichiometric methane-air mixture at various spark times.

From Figs. 12-14 it is observed that the in-cylinder pressure increases as the spark time deviates from TDC. If combustion starts too early in the cycle, the work transfer from the piston to the gases in the cylinder at the end of the compression stroke is too large. If the combustion starts too late, the peak cylinder pressure is reduced, and the stroke work transfer from the gas to the piston decreases. Another observation that is brought out from these figures is that higher pressures are obtained with methane-air mixture followed by downdraft syngas-air mixture and lastly by updraft syngas-air mixture, which represents the same behavior observed in the single compression case.

Making a parallel with the laminar combustion case where the performances of updraft and downdraft syngas are similar (see Ref. [24]). This behaviour is not found in turbulent conditions, where peak pressure of downdraft syngas is higher in about 25%. As the turbulent burning velocity could be considered as proportional to the laminar one [26], the correlations (Eq. 5, 6) shows that the laminar burning velocity increases with temperature increase and decreases with the increase of pressure. Temperature is irrelevant in this comparison since the temperature coefficient is similar for both syngas compositions. However, pressure coefficient for updraft syngas is 40% higher in relation to downdraft syngas coefficient. This means that the higher pressures used on the RCM have a higher impact in reducing the laminar burning of updraft syngas composition and, thus, justifying the lower pressures obtained in turbulent conditions.

3.2.1. *Direct light visualization from chemiluminscence emission*

Burning of a mixture in a cylinder of a SI engine may be divided into the following phases: (1) spark ignition, (2) laminar flame kernel growth and transition to turbulent combustion, (3) turbulent flame development and propagation, (4) near-wall combustion and after burning. Figures 15-17 show flame propagation images of stoichiometric syngas-air mixtures combustion and stoichiometric methane-air mixtures in a RCM, where is possible to observe these first three phases of combustion and the corresponding pressure.

In these figures, after the passage of spark, there is a point at which the beginning of pressure rise can be detected. This stage is referred to as ignition lag or preparation phase in which growth and development of a self propagating nucleus of flame takes place. This is a chemical process depending upon both pressure and temperature and the nature of the fuel. Therefore, this stage is longer for earlier ignition timings as shown in the figures 15-17. Pictures of this initial phase of combustion show an initially quasi-spherical, relatively smooth flame kernel for syngas compositions and methane.

After this stage and up to the attaining of peak pressure another stage of combustion could be considered as second stage. This is a physical one and is concerned with the spread of the flame throughout the combustion chamber. The starting point of the second stage is where the first measurable rise of pressure is seen, i.e. the point where the line of combustion departs from the compression (motoring) line. This can be seen from the deviation from the

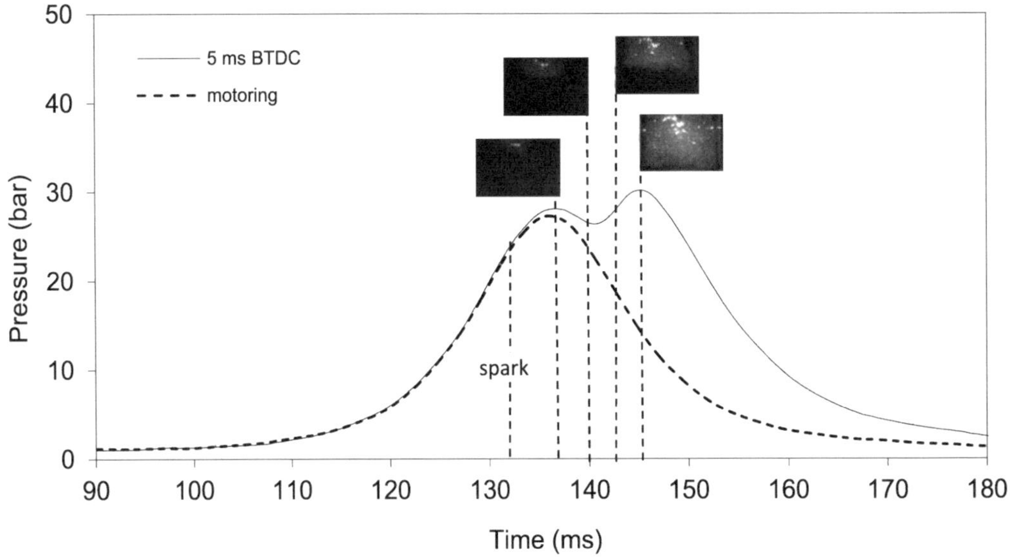

(a)

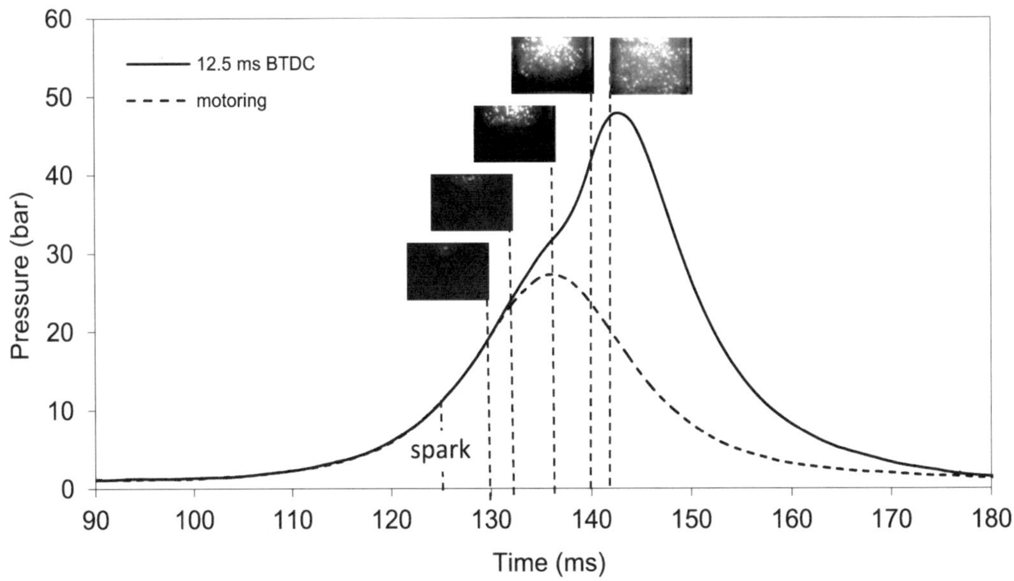

(b)

Figure 15. Direct visualization of stoichiometric updraft syngas-air flame in a RCM for various Ignition timings. (a) 5 ms BTDC; (b) 12.5 ms BTDC [25]

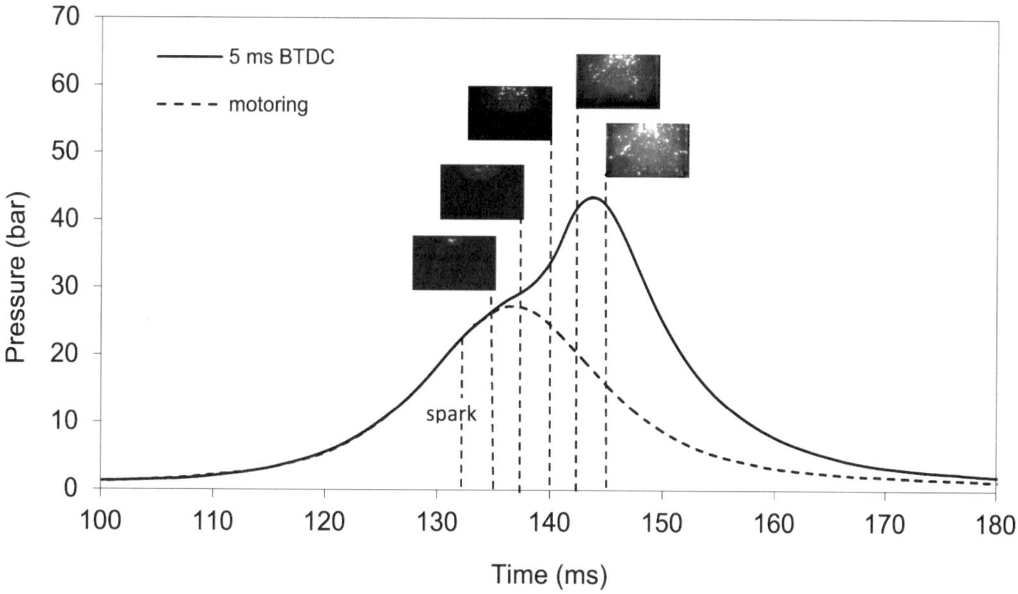

(a)

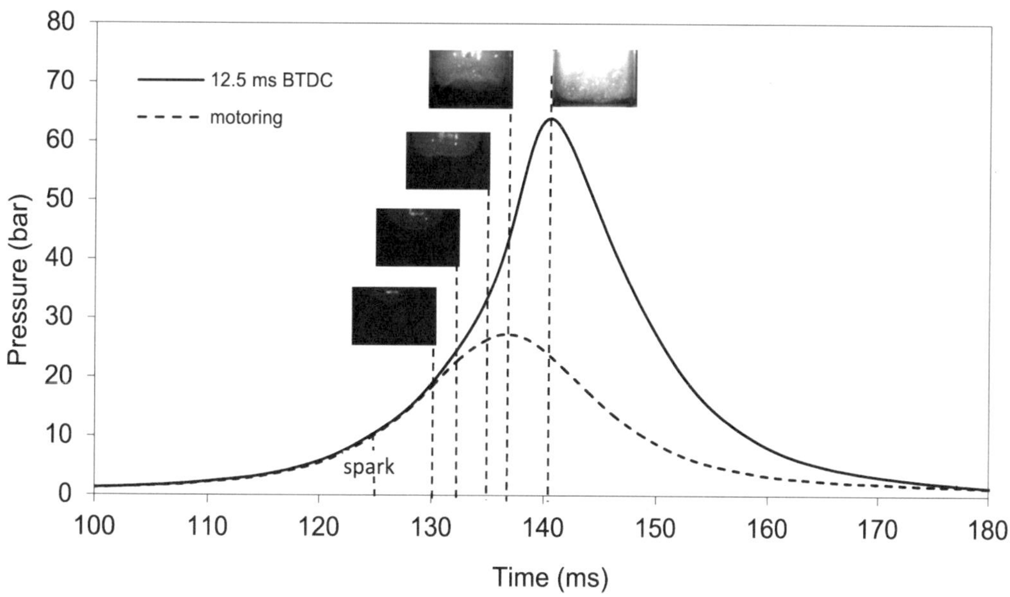

(b)

Figure 16. Direct visualization of stoichiometric downdraft syngas-air flame in a RCM for various Ignition timings. (a) 5 ms BTDC; (b) 12.5 ms BTDC [25].

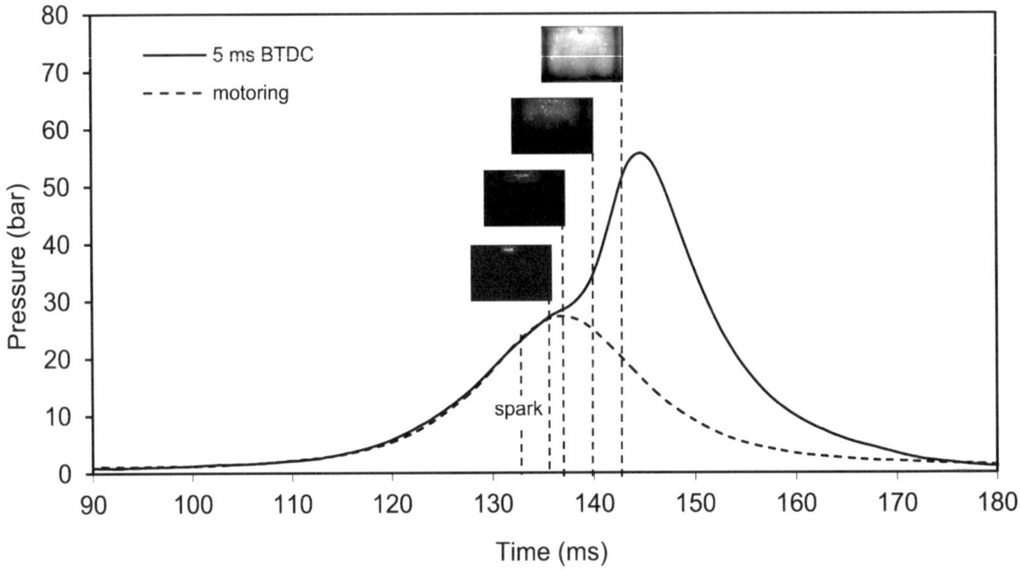

(a)

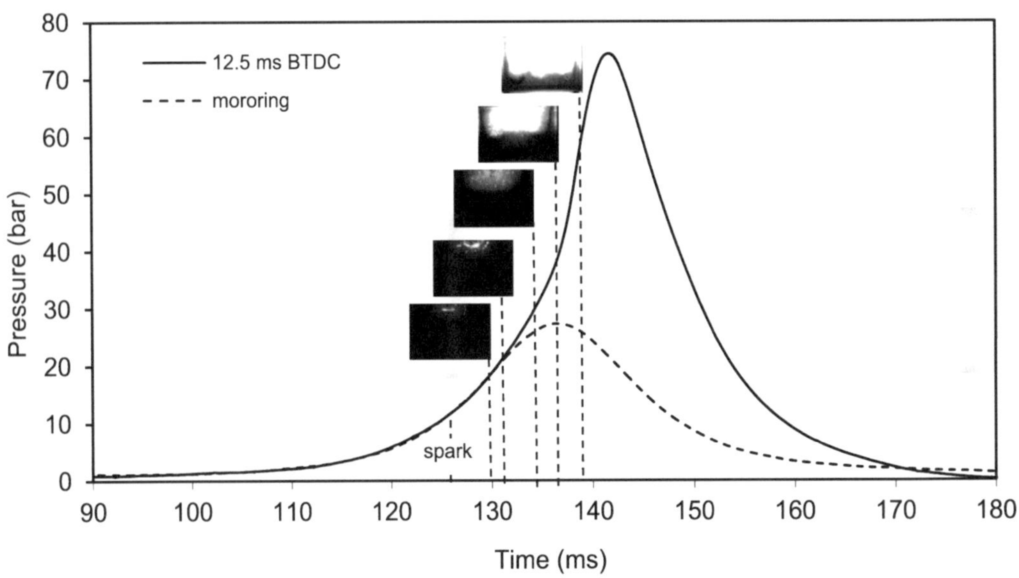

(b)

Figure 17. Direct visualization of stoichiometric methane-air flame in a RCM for various Ignition timings. (a) 5 ms BTDC; (b) 12.5 ms BTDC [25].

increase due to compression but the effect of temperature on burning velocity for typical syngas compositions is irrelevant since the temperature coefficient is of the same order. Another major finding is that syngas typical compositions are characterized by high ignition timings due to its low burning velocities.

Author details

Eliseu Monteiro and Abel Rouboa
CITAB, University of Trás-os-Montes and Alto Douro, Vila Real, Portugal

Marc Bellenoue and Julien Sottton
Institute P', ENSMA, CRNS, 86961, Futuroscope Chasseneuil Cedex, France

Acknowledgement

This work is supported by European Union Funds (FEDER/COMPETE - Operational Competitiveness Programme) and by national funds (FCT - Portuguese Foundation for Science and Technology) under the project FCOMP-01-0124-FEDER-022696.

This work was made on behalf of the FCT (Portuguese Foundation for Science and Technology) project PTDC/AAC-AMB/103119/2008.

5. References

[1] International energy agency. world energy outlook 2000, IEA, Paris, 2000.
[2] IEA bioenergy. The role of bioenergy in greenhouse gas mitigation. Position paper. IEA Bioenergy, New Zealand, 1998.
[3] A.V. Bridgwater, The technical and economic feasibility of biomass gasification for power generation, Fuel 14 (5) (1995) 631-653.
[4] C.D Rakopoulus, C.N. Michos, Development and validation of a multi-zone combustion model for performance and nitric oxide formation in syngas fueled spark ignition engine, Energy Conversion and Management, 49 (10) (2008) 2924-2938.
[5] C.R. Stone, N. Ladommatos, Design and evaluation of a fast-burn spark-ignition combustion system for gaseous fuels at high compression ratios, Journal of the Institute of Energy 64 (1991) 202–211.
[6] Li H., G.A. Karim, Experimental investigation of the knock and combustion characteristics of CH4, H2, CO, and some of their mixtures, Proceedings of the Institution of Mechanical Engineers, Part A: Journal of Power and Energy 220 (2006) 459–471.
[7] S. Rosseau, B. Lemoult, M. Tazerout, Combustion characterization of natural gas in a lean burn spark-ignition engine, Proceedings of the Institution of Mechanical Engineers, Part D: Journal of Automobile Engineering 213 (1999) 481-489.
[8] Heywood J.B., Internal combustion engine fundamentals, McGraw-Hill, New York, 1988.
[9] G. Sridhar, P.J. Paul, H.S. Mukunda, Biomass derived producer gas as a reciprocating engine fuel-an experimental analysis, Biomass Bioenergy 21 (2001) 61–72.

[10] C. Strozzi, J. Sotton, A. Mura, M. Bellenoue, Experimental and numerical study of the influence of temperature heterogeneities on self-ignition process of methane-air mixtures in a rapid compression machine, Combustion Science and Technology 180 (2008) 1829-1857.
[11] J.F. Griffiths, J.P. MacNamara, C.G.W. Sheppard, D.A.Turton, B.J. Whitaker, The relationship of knock during controlled autoignition to temperature in homogeneities and fuel reactivity, Fuel 81 (7) (2002) 2219-2225.
[12] S.M. Walton, X. He, B.T. Zigler, M.S. Wooldridge, A. Atreya, An experimental investigation of iso-octane ignition phenomena, Combust. Flame 150 (3) (2007) 246-262.
[13] G. Cho, G. Moon, D. Jeong, C. Bae, Effects of internal exhaust gas recirculation on controlled auto-ignition in a methane engine combustion, Fuel 88 (6) (2009) 1042-1048.
[14] G. Mittall, C.J. Sung, A rapid compression machine for chemical kinetics studies at elevated pressures and temperatures, Combust. Sci. Technol. 179 (2007) 497-530.
[15] C. Strozzi, Étude expérimentale de l'auto-inflammation de mélanges gazeux en milieux confines et sa modélisation avec une description cinétique chimique détaille, PhD Thesis, University of Poitiers, France, 2008.
[16] Eliseu Monteiro, Combustion study of mixtures resulting from a gasification process of forest biomass. PhD Thesis, ENSMA, France, 2011.
[17] Lewis B., von Elbe G., Combustion, Flames and Explosions of Gases, 3rd Edition, Academic Press, 1987.
[18] Metghalchi, M., and Keck, J. C., Laminar Burning velocity of propane-air mixtures at high temperatures and pressure. Combustion and Flame 38 (1980) 143-154.
[19] Ganesan V. Internal Combustion Engines. McGraw-Hill companies, 1995.
[20] Griffiths J.F, Q. Liao, A. Schreiber, J. Meyer, K.F Knoche, W. Kardylewski. Experimental and numerical studies of ditertiary butyl peroxide combustion at high pressures in a rapid compression machine. Combustion and Flame 93 (1993).303-315.
[21] Clarkson J., Griffiths J.F., Macnamara J.P., Whitaker B.J. Temperature fields during the development of combustion in a rapid compression machine. Combustion and Flame 125 (2001) 1162-1175.
[22] Minetti R., Carlier M., Ribaucour M., E. Therssen, L.R. Sochet. Comparison of oxidation and autoignition of the two primary reference fuels by rapid compression. Proceedings of the Combustion Institute 26 (1996) 747-753.
[23] Liou, T.M., Santavicca, D.A. Cycle resolved LDV measurements in a motored IC engine. Transitions of the ASME. Journal of Fluids Engineering 107 (1985) 232-240.
[24] Eliseu Monteiro, M. Bellenoue, J. Sotton, N.A. Moreira, S. Malheiro, Laminar burning velocities and Markstein numbers of syngas-air mixtures, Fuel 89 (2010) 1985-1991.
[25] Eliseu Monteiro, J. Sotton, M. Bellenoue, N. A. Moreira, S. Malheiro. Experimental study of syngas combustion at engine-like conditions in a rapid compression machine. Experimental Thermal and Fluid Sciences 35 (2011) 1473-1479.
[26] S. Verhelst, R. Sierens, A quasi-dimensional model for power cycle of a hydrogen-fuelled ICE, International Journal of Hydrogen Energy 32 (2007) 3545-3554.
[27] Alla A.G.H., Computer simulation of a four stroke spark ignition engine, Energy Conversion and Management 43 (2002) 1043-1061.
[28] J. Hartman, How to Tune and Modify Engine Management Systems, Motorbooks, 2004.

Chapter 4

Thermodynamic Study of the Working Cycle of a Direct Injection Compression Ignition Engine

Simón Fygueroa, Carlos Villamar and Olga Fygueroa

Additional information is available at the end of the chapter

http://dx.doi.org/10.5772/50028

1. Introduction

Currently, one of the worldwide most used *energy sources* are fuels derived from oil, such as hydrocarbons, that burn with oxygen releasing large amount of thermal energy. This energy can be transformed into mechanical work by mean of internal combustion engines [1]. Internal combustion engine is a device that allows obtaining mechanical energy from the thermal energy stored in a fluid due to a combustion process [2].

It should be noted that in reciprocating internal combustion engines (RICE) the combustion products constitute the working fluid; this simplifies their design and produces high thermal efficiency. For this reason these engines are one of the lighter weight generating units known and thus actually are the most commonly used transport engines [3].

The RICE operate following a *mechanical cycle* consisting of two main parts: the first one is the closed cycle, where the compression, combustion and expansion processes are carried out, and the second one is the open cycle where the working fluid is renewed, known as the gas exchange process and constituted by the intake and exhaust processes [4]. When RICE is study, it is mandatory to determine the working fluid thermodynamic properties, as well as the mixture amount that enters and leaves the cylinder [5].

The *flow characteristics* in spark-ignition engines (SIE) or compression ignition engines (CIE) can be summarized according to [6] as follows: transient as a result of piston movement, fully turbulent for all cylinders due to engine velocities and admission duct dimensions, and three-dimensional due to the engine geometry that also varies during the cycle (contours varying with time) producing different local velocity fields.

During the *gas exchange* ondulatory and inertial phenomena processes occur, as well as instability in the processes that occur within the engine. The variation of in-cylinder pressure during the intake and the exhaust has a complex pattern, for this reason the analytical calculation of gas exchange considering the above-mentioned phenomena is quite complicated and requires the use of specialized computer programs that use coefficients obtained experimentally [1].

The basis for calculation of non-stationary non-isentropic flow characteristic of RICE inlet and outlet ducts, and NO emissions was established in [7]. Various empirical correlations to take into account heat transfer during gas exchange process and to adjust the Reynolds number exponential factor in such a way that reduce to only one the adjustment coefficients were considered in[1,2,4 and 8].

A procedure widely used in both experimental and theoretical study of flow in engines, is to analyze the engine cycle in absence of combustion, simulating the process of compression-expansion and making measurements to the engine operating at this condition [5].

In all RICE working cycle processes, there is heat transfer to the cylinder walls, which occurs with greater intensity during the combustion and expansion due to the high temperature gradients reached. Woschni [9] proposed equations to determine turbulent convective heat transfer considering average speed of in-cylinder gases and Annand [10] find correlations to calculate instant average coefficients for turbulent convection heat transfer using gas average temperature and proposed correlations to evaluate flame radiation emitted during combustion. Correlations for convection heat transfer taking into consideration surface change and cylinder enclosed volume as piston moves was established in [11]. A computer program to calculate the heat transfer in a RICE combustion chamber using models to consider turbulence was presented in [12]. A universal correlation for mixture flow in the admission and exhaust process, correcting the coefficients of Nusselt, Reynolds and Prandtl numbers was proposed in [13].

In present study the *compression* process is considered adiabatic and reversible, but in real engines there is heat transfer between working fluid, valves and cylinder walls. At the beginning of compression the fluid temperature is lower than the temperature of the surfaces that surround the cylinder volume, causing an increase in the fluid temperature, some instant later temperatures get equal and latter on, heat is transferred from the working fluid to the walls, therefore the politropic coefficient varies during the process [1].

The complexity of the combustion process in RICE because of untimely and incomplete combustion, dissociation and heat transfer, has encouraged the development of special techniques for carrying out studies. Adequate realization of this process is decisive in terms of the produced power and its efficiency having great influence on the engine life and reliability [14].

Various *models* have been proposed to study the combustion process such as the Wiebe burning law applicable to SIE and the Watson law, applicable to CIE [15]. These laws determine burned mass fraction and heat released depending on the crankshaft rotation angle. These models used physical constants obtained experimentally. The Rasselier and Withrow relationship, along with burning laws allow obtaining combustion pressure per crankshaft rotation degree. To quantify *ignition delay* there are many correlations as the one proposed by [16], [17], [18] or [19]. Models proposed by [15], [20] and [17] are used to calculate the burning factor, which is representative of the burned mass fraction in the pre-mixed phase and the diffusive phase.

Volume changes may be evaluated using an expression proposed by [15], which correlates the engine dimensions: compression ratio, displaced volume, combustion chamber volume, crank radius, connecting rod length, and crankshaft rotation angle. Average temperature during combustion process can be determined using in-cylinder pressure and ideal gas equation [4].

Calculation methods used to obtain equilibrium composition and final state of chemical species present in the combustion products of a fuel air mixture are well known and are referenced in the literature [21, 22, 23 and 24]. One of the most complete programs is perhaps the NASA-Lewis code CEC [25 and 26] which considers liquid and gaseous chemical species, is extremely versatile and can be used to calculate thermodynamic state, chemical equilibrium, rockets theoretical behavior and even Chapman-Jouguet detonation properties.

Computer programs for calculating CHO and CHON constant pressure combustion systems assuming that combustion products were composed of eight and ten chemical species were presented respectively in [27 and 28]. A code less general than the NASA code, limited to twelve chemical species CHON combustion systems specifically designed to be applied to the analysis of internal combustion engines processes was published in [29]. A program for calculating twelve species CHON constant volume combustion systems, applicable to temperatures up to 3400 K was presented in [30]. A program, valid for temperatures up to 6000 K, which can be calculate, both constant pressure and constant volume combustion for an eighteen chemical species CHON system is available in [31].

The working fluid properties function of its temperature, pressure and richness can be determined by applying thermodynamic basic equations for ideal gas mixtures, considering the mass fractions of each component in the mixture [32]. Also can be determined through routines such as FARG and ECP [33 and 34]. To complement the study of the combustion process, models to determine NO emissions as the extended Zeldovich mechanism, have been considered. The reason for the use of these models is because the specific constants of the reaction rate for NO are very small compared to the combustion rate, for this reason it is supposed that all the species present in the products with the exception of NO, are in chemical equilibrium.

The *expansion* process produces mechanical work from energy released during combustion and ends when exhaust valve opens. At this moment products are expelled from the cylinder initially at critical speed ranging between 500 and 700 m/s, and then are pushed by piston movement towards the upper dead center [4 and 15]. Towards the end of the exhaust during the valve overlap, part of the fresh mixture escapes contributing to the emission of unburned hydrocarbons and reducing the engine efficiency.

To investigate the gas exchange process, using gas dynamics to analyze the gas flow in transient processes with variable composition and variable specific heats, models such as [35] have been used.

To improve the gas exchange process we must advance inlet valve opening (AIVO) and delay exhaust valve closure (DEVC). Because of this, there is a period in which the two valves remain open simultaneously, this period is known as *valve overlap*, which helps to remove as much gas and admit as much air or fresh mixture. This is due to the depression originated in the inlet valve vicinity, due to the ejection effect produced by burned gas movement through the exhaust valve; this will contribute to increase efficiency and power produced by the RICE [1].

Two research methods are employed to study the working cycle of RICE. The first one is based on data acquisition from experimental tests and the second one is based only on mathematical simulation. The latter method is more versatile and reduces the required research empirical data depending on the employed calculating method and imposed simplifications. However, to validate mathematical simulation results, experimental parameters obtained in laboratory are required [5]. Use of numerical analysis methods has now greatly developed and increased due to increase of velocity and calculation capacity of modern computers. These methods provide faster performance, versatile and can handle more information than can be measured in an experimental test. However, results accuracy obtained by applying the models depends on made assumptions.

Modeling is a research technique employed in RICE, its use has grown in last two decades due to the cost decrease obtained by eliminating or reducing the laboratory tests, since these require a large amount of repetitive tests to obtain appropriate results, bringing time and money losses in preparing, calibrating, measuring, repairing and replacing testing engines. RICE designers must build more efficient engines due to higher fuel cost and new regulations on combustion emissions produced in the process that occurs inside engine. In order to optimize these designs, numerous trial and error tests are required. Implementing the tests implies expensive construction and testing of several prototypes. Modeling is a procedure that allows realizing numerous tests with relatively low cost.

To determine engine p vs. V diagram, working fluid is considered as an ideal gas, the mass entering the cylinder is calculated using a filling model that takes into account valve rise and discharge coefficient. Initial in-cylinder mass are residual gases, same quantity that was used as a reference value to control the expelled mass during exhaust. Instant volume was determined using an equation in terms of crankshaft rotation angle [15]. Final compression temperature was found from the first law of thermodynamic considering a uniform flow process and convection heat transfer. Power, mean indicated pressure and maximum pressure and temperature were calculated using the methods proposed in [1], [4] and [15]. Cyclic dispersion were studied using mean indicated pressure variation coefficient and pressure variation as a function of main combustion phase angle in the range of 10° before top dead center (TDC) and 10° after TDC, [1]. Calculations for exhaust process were similarly to those of the admissions process. A model to study the closed loop of a CIE limited pressure cycle, replacing the constant volume heat rejection process by an isentropic expansion process followed by a constant pressure heat rejection is proposed in [36].

There are commercial packages which constitute a very useful tool in the field of research and development of RICE being employed by different companies in the automotive sector. These include the ECARD (Engine Computer Aided Research & Development), developed by the IMST group, a global model that allows simulating engine operation throughout its working cycle, using similar complexity models for the different processes involved. The OpenWAM is a free, open source 1-dimensional gas-dynamics code produced by CMT group that can be used to predict the flow movement through the elements of an internal combustion engine. NEUROPART, uses neural networks to determine the product properties and composition influence on the exhaust emissions and particles formation. CHEMKIN, uses the chemical kinetic concepts to analyze fluids in gas phase through fluid dynamic simulation. EQUIL, calculates the composition at equilibrium of combustion products. PREMIX, calculates combustion speed for different fuels. SENKIN, allows to determine the time delay for different fuels and the combustion kinetic evolution depending on the species involved in a process.

2. Mathematical model

Present paragraph will develop fundamentals and mathematical equations that govern the phenomena occurring in CIE. For this purpose, volume control in Figure 1, which shows the mass and energy interactions with surroundings will be considered.

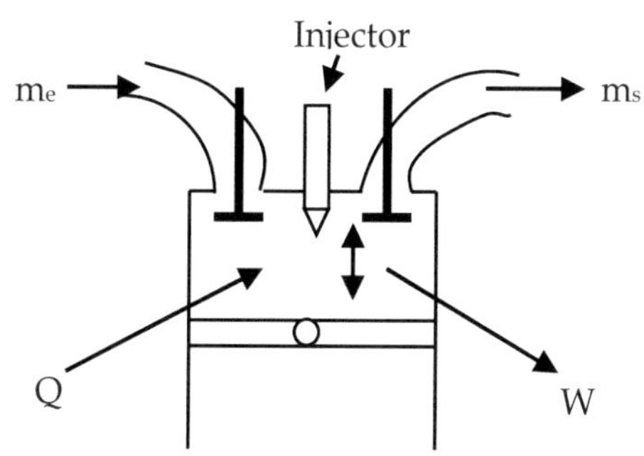

Figure 1. Engine control volume

It should be noted that the control volume during gas exchange processes works as an open system. During compression, combustion and expansion processes works as a closed system, that is why corrections should be made to take into account the exchanged mass due to leakage and the supply of fuel.

2.1. Mass conservation

The mass conservation principle establishes that the total mass change in a control volume is:

$$m_{vc} = \sum m_e - \sum m_s \tag{1}$$

The summation is used when there are several inputs and/or output flows. Expressing Ec. 1 in differential form and dividing by a time differential we obtain the mass time rate of change:

$$\frac{dm_{vc}}{dt} = \frac{dm_e}{dt} - \frac{dm_s}{dt} \tag{2}$$

To express last equation in terms of air and fuel mass entering the control volume, we define:

$$f = \frac{m_f}{m} \tag{3}$$

Time differentiating and rearranging we obtain the fuel time rate of change:

$$\dot{f} = \frac{df}{dt} = \left(\frac{\dot{m}_e - \dot{m}_s}{m} \right) \left[\left(f_e - f_s \right) \right] \tag{4}$$

From equivalence ratio (mixture richness) definition:

$$\phi = \frac{\dfrac{m_f}{m_a}}{\left(\dfrac{m_f}{m_a} \right)_{sto}} = \frac{\dfrac{\dot{m}_f}{\dot{m}_a}}{\left(\dfrac{\dot{m}_f}{\dot{m}_a} \right)_{sto}} \tag{5}$$

replacing Ec. 3 in Ec. 5 and time deriving:

$$\dot{\phi} = \frac{d\phi}{dt} = \frac{1}{\left(\dfrac{m_f}{m_a} \right)_{sto}} \frac{\dot{f}}{\left(1 - f \right)^2} \tag{6}$$

2.2. Energy conservation

The first law of thermodynamics for an open system, disregarding changes in kinetic and potential energy can be written in differential form as:

$$\frac{dE}{dt} = \frac{dQ}{dt} - \frac{dW}{dt} + \dot{m}_e h_e - \dot{m}_s h_s \tag{7}$$

Since the work due to a volume change is:

$$\frac{dW}{dt} = \dot{W} = P\frac{dV}{dt} \tag{8}$$

and the first term on the left-hand side of Eq. 7 can be evaluated in terms of the internal energy:

$$\frac{dE}{dt}=\frac{d}{dt}\left(mu\right)=\left(m\frac{du}{dt}\right)_{vc}+\left(u\frac{dm}{dt}\right)_{vc} \tag{9}$$

or in terms of enthalpy:

$$\frac{dE}{dt}=\frac{d}{dt}\left(mh\right)-\frac{d}{dt}(pV) \tag{10}$$

Substituting Eqs. 8 and 10 in Eq. 7 we have:

$$\left(m\frac{du}{dt}\right)_{vc}+\left(u\frac{dm}{dt}\right)_{vc}=\dot{Q}-\dot{W}+\dot{m}_e h_e-\dot{m}_s h_s \tag{11}$$

Since internal energy, enthalpy and density are T, p and φ functions, their time rate of change are:

$$\frac{du}{dt}=\left(\frac{\partial u}{\partial T}\right)\frac{dT}{dt}+\left(\frac{\partial u}{\partial p}\right)\frac{dp}{dt}+\left(\frac{\partial u}{\partial \phi}\right)\frac{d\phi}{dt} \tag{12}$$

$$\frac{dh}{dt}=\left(\frac{\partial h}{\partial T}\right)\frac{dT}{dt}+\left(\frac{\partial h}{\partial p}\right)\frac{dp}{dt}+\left(\frac{\partial h}{\partial \phi}\right)\frac{d\phi}{dt} \tag{13}$$

$$\frac{d\rho}{dt}=\left(\frac{\partial \rho}{\partial T}\right)\frac{dT}{dt}+\left(\frac{\partial \rho}{\partial p}\right)\frac{dp}{dt}+\left(\frac{\partial \rho}{\partial \phi}\right)\frac{d\phi}{dt} \tag{14}$$

Assuming the working fluid is an ideal gas, differentiating ideal gas equation and rearranging we have:

$$p\frac{dV}{dT}+V\frac{dp}{dT}=mR\frac{dT}{dt}+mT\frac{dR}{dt}+RT\frac{dm}{dt} \tag{15}$$

$$\frac{d\rho}{dt}=\frac{\frac{dp}{dt}-\rho R\frac{dT}{dt}-\rho T\frac{dR}{dt}}{RT} \tag{16}$$

From Eq. 14:

$$\frac{dp}{dt}=\frac{\frac{d\rho}{dt}-\left(\frac{\partial \rho}{\partial T}\right)\frac{dT}{dt}-\left(\frac{\partial \rho}{\partial \phi}\right)\frac{d\phi}{dt}}{\left(\frac{\partial \rho}{\partial p}\right)} \tag{17}$$

substituting Eq. 17 in Eq. 16, rearranging and solving for $\frac{dp}{dT}$:

$$\frac{dp}{dt}=\frac{-\frac{\rho}{T}\frac{dT}{dt}-\frac{\rho}{R}\frac{dR}{dt}-\left(\frac{\partial\rho}{\partial T}\right)\frac{dT}{dt}-\left(\frac{\partial\rho}{\partial\phi}\right)\frac{d\phi}{dt}}{\left(\frac{\partial\rho}{\partial p}\right)-\frac{1}{RT}} \tag{18}$$

Solving Ec. 15 for $\frac{dR}{dt}$, simplifying and substituting in Ec. 18:

$$\frac{dp}{dt}\left[\left(\frac{\partial\rho}{\partial p}\right)-\frac{1}{RT}\right]=-\frac{\rho}{T}\frac{dT}{dt}-\frac{\rho}{R}\left[\frac{p}{mT}\frac{dV}{dt}+\frac{V}{mT}\frac{dp}{dt}-\frac{R}{T}\frac{dT}{dt}-\frac{R}{m}\frac{dm}{dt}\right]-\left(\frac{\partial\rho}{\partial T}\right)\frac{dT}{dt}-\left(\frac{\partial\rho}{\partial\phi}\right)\frac{d\phi}{dt} \tag{19}$$

Replacing ideal gas equation in Ec. 19 and solving for $\frac{dp}{dT}$:

$$\frac{dp}{dt}=\frac{\rho}{\left(\frac{\partial\rho}{\partial p}\right)}\left[\frac{\frac{dV}{dt}}{V}+\frac{\frac{dm}{dt}}{m}-\frac{1}{\rho}\left(\frac{\partial\rho}{\partial T}\right)\frac{dT}{dt}-\frac{1}{\rho}\left(\frac{\partial\rho}{\partial\phi}\right)\frac{d\phi}{dt}\right] \tag{20}$$

Differentiating ideal gas equation with respect to p and T, we obtain:

$$\left(\frac{\partial\rho}{\partial p}\right)=\frac{1}{RT} \tag{21}$$

$$\left(\frac{\partial\rho}{\partial T}\right)=-\frac{p}{RT^2} \tag{22}$$

and substituting Ecs. 21 and 22 in Ec. 20:

$$\frac{dp}{dt}=p\left[-\frac{\frac{dV}{dt}}{V}+\frac{\frac{dm}{dt}}{m}+\left(\frac{dT}{dt}\right)\frac{1}{T}-\frac{RT}{p}\left(\frac{\partial\rho}{\partial\phi}\right)\frac{d\phi}{dt}\right] \tag{23}$$

Later equation, function of density, volume, mass and mixture richness will be used to obtain the in-cylinder pressure when time varies (indicator diagram).

The procedure to obtain a similar expression for temperature variation over time will be illustrated below. Solving Eq. 11 for $\frac{du}{dt}$:

$$\frac{du}{dt}=\frac{\dot{Q}}{m_{vc}}-\frac{p}{m_{vc}}\frac{dV}{dt}+\frac{1}{m_{vc}}\left(\dot{m}_e h_e-\dot{m}_s h_s-\left(u\frac{dm}{dt}\right)_{vc}\right) \tag{24}$$

and defining:

$$B = -RT\frac{\frac{dV}{dt}}{V} + \frac{1}{m}\left(\dot{Q} + \dot{m}_e h_e - \dot{m}_s h_s - \left(u\frac{dm}{dt}\right)_{vc}\right) \tag{25}$$

On the other hand, introducing Ecs. 8 in Ec. 11, get the following expression:

$$\frac{du}{dt} = B \tag{26}$$

Substituting Ec. 26 in Ec. 12 and solving for $\frac{dp}{dt}$:

$$\frac{dp}{dt} = \frac{B - \left(\frac{\partial u}{dt}\right)\frac{dT}{dt} - \left(\frac{\partial u}{d\phi}\right)\frac{d\phi}{dt}}{\left(\frac{\partial u}{\partial p}\right)} \tag{27}$$

Replacing Ec. 27 in Ec. 20 and solving for $\frac{dT}{dt}$ gives:

$$\frac{dT}{dt} = \frac{\left(\frac{\partial u}{\partial p}\right)\left[-\rho\frac{\frac{dV}{dt}}{V} + \rho\frac{\frac{dm}{dt}}{m} - \left(\frac{\partial \rho}{\partial \phi}\right)\frac{d\phi}{dt} - B + \left(\frac{\partial u}{\partial \phi}\right)\frac{d\phi}{dt}\right]}{\left(\frac{\partial u}{\partial p}\right)\left(\frac{\partial \rho}{\partial T}\right) - \left(\frac{\partial \rho}{\partial p}\right)\left(\frac{\partial u}{\partial T}\right)} \tag{28}$$

Now, considering:

$$R = R(T, p, \phi, t) \tag{29}$$

And differentiating:

$$\frac{dR}{dt} = \left(\frac{\partial R}{\partial T}\right)\frac{dT}{dt} + \left(\frac{\partial R}{\partial p}\right)\frac{dp}{dt} + \left(\frac{\partial R}{\partial \phi}\right)\frac{d\phi}{dt} \tag{30}$$

Differentiating ideal gas equation and solving for $\frac{dR}{dt}$

$$\frac{dR}{dt} = \frac{\frac{dp}{dt}}{p}R - \frac{\frac{dT}{dt}}{T}R - \frac{\frac{d\rho}{dt}}{\rho}R \tag{31}$$

Replacing Ec. 30 in Eq. 31 and solving for $\frac{dp}{dt}$:

$$\frac{dp}{dt}=\frac{\left(\frac{\partial R}{\partial T}\right)\frac{dT}{dt}+\left(\frac{\partial R}{\partial \phi}\right)\frac{d\phi}{dt}+\frac{dT}{dt}\frac{R}{T}+\frac{d\rho}{dt}\frac{R}{\rho}}{\frac{R}{p}-\left(\frac{\partial R}{\partial p}\right)} \tag{32}$$

Replacing Ec. 32 in Ec. 27:

$$\left(\frac{\partial u}{\partial T}\right)\frac{dT}{dt}+\left(\frac{\partial u}{\partial p}\right)\left[\frac{\left(\frac{\partial R}{\partial T}\right)\frac{dT}{dt}+\left(\frac{\partial R}{\partial \phi}\right)\frac{d\phi}{dt}+\frac{dT}{dt}\frac{R}{T}+\frac{d\rho}{dt}\frac{R}{\rho}}{\frac{R}{p}-\left(\frac{\partial R}{\partial p}\right)}\right]+\left(\frac{\partial u}{\partial \phi}\right)\frac{d\phi}{dt}=B \tag{33}$$

Defining:

$$D=1-\frac{p}{R}\left(\frac{\partial R}{\partial p}\right) \tag{34}$$

Collecting terms containing $\frac{dT}{dt}$ and substituting Eq. 34 in Eq. 33:

$$\left(\frac{dT}{dt}\right)\left[\left(\frac{\partial u}{\partial T}\right)+\left(\frac{\partial u}{\partial p}\right)\frac{p}{DR}\left\{\left(\frac{\partial R}{\partial T}+\frac{R}{T}\right)\right\}\right]+\left(\frac{\partial u}{\partial p}\right)\frac{1}{D}\left[\left(\frac{\partial R}{\partial \phi}\right)\frac{d\phi}{dt}+\frac{d\phi}{dt}\frac{R}{\rho}\right]+\left(\frac{\partial u}{\partial \phi}\right)\frac{d\phi}{dt}=B \tag{35}$$

Defining:

$$C=1+\frac{T}{R}\left(\frac{\partial R}{\partial T}\right) \tag{36}$$

Replacing Ec. 36 in Eq. 35 and solving for $\frac{dT}{dt}$:

$$\frac{dT}{dt}=\frac{B-\left(\frac{\partial u}{\partial p}\right)\frac{p}{D}\left[\frac{1}{R}\left(\frac{\partial R}{\partial \phi}\right)\frac{d\phi}{dt}+\frac{d\rho}{dt}\frac{1}{\rho}\right]-\left(\frac{\partial u}{\partial \phi}\right)\frac{d\phi}{dt}}{\left(\frac{\partial u}{\partial T}\right)+\frac{C}{D}\frac{p}{T}\left(\frac{\partial u}{\partial p}\right)} \tag{37}$$

Since:

$$m=\rho V \tag{38}$$

Time differentiating and solving for $\frac{d\rho}{dt}$:

$$\frac{d\rho}{dt}=\frac{\frac{dm}{dt}-\rho\frac{dV}{dt}}{V} \tag{39}$$

Replacing Ec. 39 in Ec. 37:

$$\frac{dT}{dt}=\frac{B-\left(\frac{\partial u}{\partial p}\right)\frac{p}{D}\left[\left(\frac{\partial R}{\partial \phi}\right)\frac{d\phi}{dt}\frac{1}{R}+\frac{dm}{dt}\frac{1}{m}-\frac{dV}{dt}\frac{1}{V}\right]-\left(\frac{\partial u}{\partial \phi}\right)\frac{d\phi}{dt}}{\left(\frac{\partial u}{\partial T}\right)+\frac{C}{D}\frac{p}{T}\left(\frac{\partial u}{\partial p}\right)} \tag{40}$$

This equation will be used to determine the in-cylinder temperature when time varies.

If Eqs. 25, 34, 36 and 37 are replaced in Eq. 23 and collecting terms we obtained:

$$\frac{dp}{dt}=\frac{\dot{Q}-\dot{m}_{bb}h_{bb}-\frac{dV}{dt}\left[\frac{mC}{V}+p\right]-\frac{dm}{dt}\left[\left(D\left(\frac{\partial u}{\partial \phi}\right)-h_{cil}+u\right)-C\left(\frac{D}{R}\left(\frac{\partial R}{\partial \phi}+1\right)\right)\right]}{m\left[C\left(\frac{1}{p}-\frac{1}{R}\left(\frac{\partial R}{\partial \phi}\right)\right)+\left(\frac{\partial u}{\partial p}\right)\right]} \tag{41}$$

and:

$$\frac{dm}{dt}=\frac{\dot{Q}-\dot{m}_{bb}h_{bb}-\frac{dV}{dt}\left[\frac{mC}{V}+p\right]-m\left[C\left(\frac{1}{p}-\frac{1}{R}\left(\frac{\partial R}{\partial \phi}\right)\right)+\left(\frac{\partial u}{\partial p}\right)\right]\frac{dp}{dt}}{\left(D\left(\frac{\partial u}{\partial \phi}\right)-h_{cil}+u\right)-C\left(\frac{D}{R}\left(\frac{\partial R}{\partial \phi}+1\right)\right)} \tag{42}$$

Eqs. 41 and 42 will be used to obtain the indicator diagram (p vs. V or p vs. ϕ diagram) and burned mass fraction diagram (m vs. t diagram) respectively.

2.3. Instant in-cylinder volume

The instant volume inside the control volume in terms of the displaced volume, compression ratio, connecting rod length to crank radius ratio and crankshaft rotation angle, can be obtained through the following expression [15]:

$$V(\varphi)=V_d\left[\frac{1}{r_c-1}+\frac{1}{2}\left[R_{LA}+1-\cos\varphi-(R_{LA}^2-sen^2\varphi)^{\frac{1}{2}}\right]\right] \tag{43}$$

In this expression R_{LA} is the connecting rod length (l) to crank radius (a) ratio.

$$R_{LA}=\frac{l}{a} \tag{44}$$

Deriving Ec. 43 with respect to crankshaft rotation angle, we obtain:

$$\frac{dV}{d\varphi} = \frac{V_d}{2}\left[sen\varphi + \frac{sen\varphi\ cos\varphi}{\left(R_{LA}^2 - sen^2\varphi\right)^{\frac{1}{2}}}\right] \tag{45}$$

The time in seconds it takes to describe some crankshaft rotation angle can be calculated with the following expression:

$$t = \frac{\varphi}{6\ rpm} \tag{46}$$

Solving previous expression for ф and replacing in Ec. 45 to make the corresponding conversion from degrees to radians we obtained:

$$\frac{dV}{dt} = 3V_d(rpm)\left[sen\frac{\pi\ rpm}{30}t + \frac{sen\frac{\pi\ rpm}{30}t\ cos\frac{\pi\ rpm}{30}t}{\left(R_{LA}^2 - sen^2\frac{\pi\ rpm}{30}t\right)^{\frac{1}{2}}}\right] \tag{47}$$

Previous expression allows determining the in-cylinder volume variation with respect to time, while Ec. 45 will be used to calculate the volume variation with respect to crankshaft rotation angle.

3. Equations, models and calculations

Models and assumptions used to analyze each of the thermodynamic processes that are carried out in a CIE will be presented in this paragraph. Routines commonly used in RICE are employed to calculate the thermodynamic properties of the chemical species formed during combustion. FARG and ECP routines [34] are used to determine properties depending on gas temperature. PER routine [29] is used to obtain same properties depending on mixture richness. DVERK routine [37] found in the International Mathematics and Statistics Library is used for solving differential equations systems by the Runge - Kutta Verner fifth and sixth order method.

3.1. Admission process

The parameter characterizing the admissions process is the volumetric efficiency defined as:

$$\eta_v = \frac{\dot{m}_{ar}}{\dot{m}_{at}} = \frac{\dot{m}_{ar}}{\rho_0 i V_d \frac{rpm}{30j}} \tag{48}$$

It takes into account the losses in the inlet valve and all the admission system if atmospheric density value is used for ϱ_0.

Real mass air flow entering the cylinder is determined by the following equations [15] function of the ratio p_{down}/p_{up}:

$$\frac{p_{down}}{p_{up}} < 1 \qquad \dot{m} = \frac{C_d A_{ref} p_0}{\sqrt{R_a T_0}} \left(\frac{p_{down}}{p_{up}}\right) \left(\frac{2\gamma}{\gamma - 1} \left[1 - \left(\frac{p_{down}}{p_{up}} \right)^{\frac{\gamma+1}{\gamma}} \right] \right) \tag{49}$$

$$\frac{p_{down}}{p_{up}} \geq 1 \qquad \dot{m} = C_d A_{ref} p_{up} \sqrt{ \left(\frac{\gamma}{RT_0} \right) \left(\frac{2}{\gamma - 1} \right)^{\frac{\gamma+1}{2(\gamma+1)}} } \tag{50}$$

where p_{down} is the downstream pressure and p_{up} is the upstream pressure. Although the discharge coefficient C_d varies during the process, in present study, we assume that it is constant and equal to its average value. The reference area A_{ref}, usually called curtain area, since it depends on the valve lifting L_v, is taken as:

$$A_{ref} = \pi d_v L_v \tag{51}$$

A model proposed in [38], was used to theoretically determine the lifting valve profile which is function of the maximum lifting, $L_{v\,max}$ and the crankshaft rotation angle ϕ:

$$L_v(\varphi) = L_{v\max} + C_2 \varphi^2 + C_p \varphi^p + C_q \varphi^q + C_r \varphi^r + C_s \varphi^s \tag{52}$$

Coefficients C_2, C_p, C_q, C_r y C_s are determined with the following equations:

$$C_2 = \frac{-pqrsh}{\left[(p-2)(q-2)(r-2)(s-2)c_{med}^2\right]} \tag{53}$$

$$C_p = \frac{2qrsh}{\left[(p-2)(q-p)(r-p)(s-p)c_{med}^p\right]} \tag{54}$$

$$C_q = \frac{-2prsh}{\left[(q-2)(q-p)(r-q)(s-q)c_{med}^q\right]} \tag{55}$$

$$C_r = \frac{2pqsh}{\left[(r-2)(r-p)(r-q)(s-r)c_{med}^r\right]} \tag{56}$$

$$C_s = \frac{-2pqrh}{\left[(s-2)(s-p)(s-q)(s-r)c_{med}^s\right]} \tag{57}$$

Recommended values for p, q, r, and s are: p = 6; q = 8; r = 10; s = 12.

Gas pressure and temperature variation over time in this process is calculated from Eqs. 23 and 40. Since a CIE only compresses air, the term corresponding to mixture richness variation with time is zero. For this reason the above equations are:

$$\frac{dp}{dt} = p\left[-\frac{\frac{dV}{dt}}{V} + \frac{\frac{dm}{dt}}{m} + \left(\frac{dT}{dt}\right)\frac{1}{T}\right] \tag{58}$$

$$\frac{dT}{dt} = \frac{B - \left(\frac{\partial u}{\partial p}\right)\frac{p}{D}\left[\frac{dm}{dt}\frac{1}{m} - \frac{dV}{dt}\frac{1}{V}\right]}{\left(\frac{\partial u}{\partial T}\right) + \frac{C}{D}\frac{p}{T}\left(\frac{\partial u}{\partial p}\right)} \tag{59}$$

To solve these equations heat release, heat transfer, blow by, ignition delay and chemical species formation models are required. In addition the terms $\frac{\partial u}{\partial T}$, $\frac{\partial u}{\partial p}$, $\frac{\partial R}{\partial T}$ y $\frac{\partial R}{\partial p}$ must be determined by using routines FARG and ECP.

With Eqs. 43 and 47 we calculate the volume and its time derivative, respectively, while with Eqs. 49 or 50 depending on the case determine the flow mass. The in-cylinder accumulated mass is obtained by summation of the mass flows times the values obtained by Eq. 46.

3.2. Closed loop cycle

Closed loop cycle corresponds to compression, combustion and expansion processes. Compression starts when inlet valve closes. Variation in pressure and temperature with time during this process is determined taking into account that only air is compressed (Ecs 58 and 59) and there are mass losses due to blow by. When fuel injection begins the mixture composition varies; therefore the expressions used to determine temperature and pressure time variation during the closed loop cycle are Eq. 23 and 40. When exhaust valve opens, begins the exhaust process.

3.3. Exhaust process

Equations used in this process are the same used during the admissions process, but noting the working fluid is a mixture of burned gases and heat transfer is higher than during admission because of the high temperature present.

During valve overlap we want to extract as much as possible of burned gases and taking advantage of the dynamic effects, increase the amount of fresh charge entering the cylinder. Equations used during this process are the same used during the intake and exhaust, but considering that there is simultaneously fresh charge entry and burned gases exit.

3.4. Ignition delay model

Ignition delay in CIE characterizes the heat quantity will release immediately occur the fuel auto ignition and has a direct influence on engine rumble and pollutants formation. The model presented in [19] points out that ignition delay depends on in-cylinder temperature and pressure, engine speed and accumulated fuel amount and may be calculated in degrees and in milliseconds with the following expressions:

$$ID_{[\deg]} = Ap_c^n(EA) \tag{60}$$

$$ID_{[ms]} = \frac{ID_{[\deg]}}{0.006(rpm)} \tag{61}$$

where: $A=0.36+0.22V_{mp}$, $V_{mp}=\frac{c(rpm)}{30}$, $n=0$, $EA=\exp\left[R_u\left(\frac{1}{RT_c}-\frac{1}{17190}\right)\left(\frac{21.2}{p_c-12.4}\right)\right]$,

$E=\frac{618840}{NC+25}$, $p_c=p_{amb}r_c^{n_c}$, $T_c=T_{amb}r_c^{n_c-1}$, $n_c=1.30$ a 1.37, $R_u=8.3143\left[J/molK\right]$.

Other models based on experimental data suggest correlations that use an Arrhenius expression similar to that proposed in [15], in which the constants estimated by [39] are as follows: $A=3.45$, $n=1.02$, $EA=\exp\left[\frac{E_a}{R_uT_c}\right]$, $\frac{E}{R_u}=2100$.

Another model whose constants are the same as in previous case use an A term function of richness as shown by the following expression [16]: $A=2.4\phi^{-0.2}$.

3.5. Heat release model

Considering the fourth term numerator in Eq. 41, which represents the heat released during the combustion process and applying the Watson relationship we obtain the following equation:

$$m_cH_idX_b = \frac{dm}{dt}\left[\left(D\left(\frac{\partial u}{\partial \phi}\right)-h_{cil}+u\right)-C\left(\frac{D}{R}\left(\frac{\partial R}{\partial \phi}+1\right)\right)\right] \tag{62}$$

Model of fuel apparent burning will be used to represent the combustion process. It uses two empirical equations, one for the pre-mixed combustion phase and another for the diffusive combustion phase. The instantaneous total amount of heat released by crankshaft rotation degree is given by the sum of the two components:

$$\left(\frac{dm_c}{d\varphi}\right)_{Tot} = \left(\frac{dm_c}{d\varphi}\right)_{pre} + \left(\frac{dm_c}{d\varphi}\right)_{dif} \tag{63}$$

3.6. Burning factor

Heat release model requires defining, depending on the process physical condition, the initial fuel amount burned during the pre-mixed phase. An initial fuel burning factor is used for this purpose [15] [17]. This factor estimated, depending on initial richness and delay period, what part of the injected fuel is burned during the pre-mixed phase. The difference is burned during the diffusive phase. The burning factor is defined as:

$$\beta = \frac{(m_c)_{pre}}{(m_c)_{Tot}} \tag{64}$$

and may be calculated by the following expression [15]:

$$\beta = 1 - \frac{a_1 \phi_0^{b_1}}{ID_S^{cc_1}} \tag{65}$$

The a1, b1 and cc1 values which are shown in Table 1 [39, 19 and 15] depend on the used model.

Value	Hardenberg model	Watson model	Heywood model
a_1	0.746	0.926	0.80 - 0.95
b_1	0.35	0.37	0.25 - 0.45
cc_1	0.35	0.26	0.25 – 0.50

Table 1. Empirical values for burning factor

Taking into account the heat released during each phase Ec. 63 becomes

$$\left(\frac{dm_c}{d\varphi}\right)_{Tot} = \beta\left(\frac{dm_c}{d\varphi}\right)_{pre} + (1-\beta)\left(\frac{dm_c}{d\varphi}\right)_{dif} \tag{66}$$

Heat released during each phase is evaluated by the empirical expressions proposed in [39] and [40]. Equations proposed in [39] are:

$$\left(\frac{dX_b}{d\varphi}\right)_{pre} = C_1 C_2 \left(\frac{\varphi - \varphi_0}{\Delta\varphi}\right)^{C_1 - 1} \left(1 - \left(\frac{\varphi - \varphi_0}{\Delta\varphi}\right)^{C_1}\right)^{C_2 - 1} \tag{67}$$

$$\left(\frac{dX_b}{d\varphi}\right)_{dif} = C_3 C_4 \left(\frac{\varphi - \varphi_0}{\Delta\varphi}\right)^{C_4 - 1} \exp\left(C_3 - \left(\frac{\varphi - \varphi_0}{\Delta\varphi}\right)^{C_4}\right) \tag{68}$$

Equation proposed in [40] that use the duration and heat released percentage in each phase, unlike [39] equations, is:

$$\frac{dQ}{d\varphi}=a\left(\frac{Q_{pre}}{\varphi_{pre}}\right)m_{pre}\left(\frac{\varphi}{\varphi_{pre}}\right)^{m_{pre}-1}\exp\left[-a\left(\frac{\varphi}{\varphi_{pre}}\right)^{m_{pre}}\right]+\left(\frac{Q_{dif}}{\varphi_{dif}}\right)m_{dif}\left(\frac{\varphi}{\varphi_{dif}}\right)^{m_{dif}-1}\exp\left[-a\left(\frac{\varphi}{\varphi_{dif}}\right)^{m_{dif}}\right] \quad (69)$$

Table 2 shows the constants for Heywood [15] and Watson [39] equations and Table 3 shows the constants for Miyamoto [40] equation.

Constant	Heywood	Watson	
		Expresión	Valores sugeridos
C_1	$2+1.25E-8\left(ID_{[ms]}rpm\right)^{2.4}$	$2+0.002703*IDa^{2.4}$	3
C_2	5000	5000	5000
C_3	$14.2\phi^{-0.644}$	$10\phi^{1.505}$	6.908
C_4	$0.79C_3^{0.25}$	$0.48C_3^{0.423}$	1.4

Table 2. Constants for Heywood and Watson expressions

Constant	Miyamoto
m_{pre}	4
m_{dif}	1.5 ID o 1.9 ID
a	0.69
ϕ_{pre}	+7°

Table 3. Constants for Miyamoto expression

3.7. Blow by model

Due to the in-cylinder pressure increase during compression and combustion processes, a part of the gasses (m_{bb}) is lost through the rings resulting in a produced power reduction. The model that will be described below takes into account such losses.

From logarithmic derivative of the ideal gas equation:

$$\frac{1}{p}\frac{dp}{d\varphi}+\frac{1}{V}\frac{dV}{d\varphi}=\frac{1}{m}\frac{dm}{d\varphi}+\frac{1}{T}\frac{dT}{d\varphi} \quad (70)$$

Applying the First Law of Thermodynamics in differential form to an open system, we obtain:

$$mC_v\frac{dT}{d\varphi}+C_vT\frac{dm}{d\varphi}=\frac{dQ}{d\varphi}-p\frac{dV}{d\varphi}-\frac{\dot{m}_{bb}C_pT}{\omega} \tag{71}$$

Replacing $dT/d\phi$ from Ec. 70 and rearranging:

$$\frac{dp}{d\varphi}=-\gamma\frac{p}{V}\frac{dV}{d\varphi}+\frac{\gamma-1}{V}\frac{dQ}{d\varphi}-\frac{\gamma\dot{m}_{bb}}{\omega m}p \tag{72}$$

Blow by mass can be found from mass conservation:

$$\frac{dm}{d\varphi}=-\frac{\dot{m}_{bb}}{\omega} \tag{73}$$

Defining:

$$C=\frac{\dot{m}_{bb}}{m} \tag{74}$$

Net heat entering the system, considering that a part is lost by blow by, is:

$$dQ=Q_e dx-dQ_{lost} \tag{75}$$

with:

$$\frac{dQ_{lost}}{dt}=h_c A(T-T_{wall}) \tag{76}$$

and defining the following dimensionless variables:

$$\tilde{p}=\frac{p}{p_1} \tag{77}$$

$$\tilde{V}=\frac{V}{V_1} \tag{78}$$

$$\tilde{T}=\frac{T}{T_1} \tag{79}$$

$$\tilde{Q}=\frac{Q_e}{P_1V_1} \tag{80}$$

$$\tilde{Q}_{lost}=\frac{Q_{lost}}{P_1V_1} \tag{81}$$

$$\tilde{h}_c = \frac{hT_1\left(A_{cc} - \frac{4V_{cc}}{D_p}\right)}{P_1V_1\omega} \tag{82}$$

$$\beta = \frac{4V_1}{D_p\left(A_{cc} - \frac{4V_{cc}}{D_p}\right)} \tag{83}$$

dimensionless heat losses for unit crankshaft rotation angle are:

$$\frac{d\tilde{Q}_{lost}}{d\varphi} = \tilde{h}_c\left(1+\beta\tilde{V}\right)\left(\tilde{p}\tilde{V} - \tilde{T}_{wall}\right) \tag{84}$$

Replacing Ecs. 77 to 83 in Ecs. 72, 73 and 84 and applying expansion work definition, we have the system of equations:

$$\frac{d\tilde{p}}{d\varphi} = -\gamma\frac{\tilde{p}}{\tilde{V}}\frac{d\tilde{V}}{d\varphi} + \frac{\gamma-1}{\tilde{V}}\left[\tilde{Q}\frac{dx}{d\varphi} - \tilde{h}_c\left(1+\beta\tilde{V}\right)\left(\frac{\tilde{p}\tilde{V}}{\tilde{m}} - \tilde{T}_{wall}\right)\right] - \frac{\gamma C\tilde{p}}{\omega} \tag{85}$$

$$\frac{d\tilde{Q}_{lost}}{d\varphi} = \tilde{h}_c\left(1+\beta\tilde{V}\right)\left(\frac{\tilde{p}\tilde{V}}{\tilde{m}} - \tilde{T}_{wall}\right) \tag{86}$$

$$\frac{dm}{d\varphi} = -\frac{Cm}{\omega} \tag{87}$$

$$\frac{d\tilde{W}}{d\varphi} = \tilde{p}\frac{d\tilde{V}}{d\varphi} \tag{88}$$

Solving this ordinary differential equations system one can calculate the mass lost by leaks, lost heat and power, and pressure variation.

In-cylinder mass varies with time and can be calculated depending on the speed of the engine. Replacing Ec. 73 in Eq. 87 and solving for $\overline{m}$ we obtain:

$$\overline{m} = \exp\left[\frac{-C_{bb}(\varphi+\pi)}{\frac{\pi(rpm)}{30}}\right] \tag{89}$$

3.8. Heat transfer models

Calculation of instant heat transfer in RICE is a complex problem. Expressions for calculating the total flow heat by combined conduction, convection and radiation requires to use empirical correlations [41]. One of the more used is based on the relationship between the Nusselt number and the Reynolds number for forced convection:

$$Nu = a\,\mathrm{Re}^{b} \tag{90}$$

Replacing the Nusselt and Reynolds numbers in Eq. 112, we have:

$$\frac{\bar{h}_c L}{k} = a\left(\frac{\rho V L}{\mu}\right)^{b} \tag{91}$$

In this expression L represents the characteristic length, which in RICE is the piston diameter, a magnitude that does not vary.

Considering also separately the radiation heat transfer, gets the expression to evaluate the total heat flow [41]:

$$\frac{q}{A} = a\frac{k}{D_p}\mathrm{Re}^{b}\,\Delta T + c\left(T_{gas}^{4} - T_{wall}^{4}\right) \tag{92}$$

In these expressions a, b and c are constants that usually take the following values: a = 0.35 – 0.80, b = 0.70 – 0.80 and c = 0.576 σ, where σ is the Stefan – Boltzmann constant.

An expression used to calculate convection heat transferred is based on the model proposed in [18]:

$$Q_{lost} = \bar{h}_c * A * \left(\bar{T}_{gas} - \bar{T}_{wall}\right) \tag{93}$$

where $\bar{h}_c$ represents the overall convection heat transfer coefficient that can be determined from the expression:

$$\bar{h}_c = a * D_p^{m-1} p^{m} T_{gas}^{0.75-1.62m} w^{m} \tag{94}$$

The constant values in this equation are: a = 0.13 and m = 0.8. Pressure must be in bar and the temperature in kelvin.

w is the average gas speed calculated by the following expression:

$$w = \left[C_{1w} V_{mp} + C_{2w} \frac{V_d T_{ref}}{p_{ref} V_{ref}} \left(p - p_{no\,comb}\right)\right] \tag{95}$$

Subscript ref is used for a reference state that may be the compression process beginning. The empirical constant values depend on the process are: C_{1w} = 6.18 during gas exchange process, C_{1w} = 2.28 during compression, combustion and expansion processes, C_{2w} = 0.0 during gas exchange and compression processes and C_{2w} = 3.24E-3 during combustion and expansion processes.

Another expression used to consider the convection heat transfer process of RICE proposed in [19] is:

$$\bar{h}_c = C_{1H} V_d^{-0.06} p^{0.8} T^{-0.4} (V_{mp} + C_{2H})^{0.8} \quad (96)$$

where C_{1H} and C_{2H} are empirical constants which take into account local variations due to intake turbulence or combustion chamber geometry of the, their values are: C_{1H} = 130E-4 and C_{2H} = 1.40. Eq. 96 considers the increase in gas speed due to engine velocity and uses as characteristic length the volume rather than the piston diameter, as it is the case when using expressions proposed in [18].

3.9. In-cylinder mass

Total mass that fills the cylinder is composed of air, fuel and residual gases. Fuel enters the cylinder when compression process end and fuel injection begins. Residual gases have two components: gases remaining in the cylinder at the end of the exhaust process and recirculated gases entering the cylinder with admitted air as a pollution control measure. There is also a mass lost by blow by. Therefore we have:

$$m_{total} = m_{aire} + m_{gr} + m_f + m_{EGR} - m_{bb} \quad (97)$$

The theoretical incoming air mass is calculated by:

$$m_{aire} = \rho V_d \quad (98)$$

While the real mass entering the cylinder is obtained by the expression:

$$m_{aire} = \sum \dot{m}_{adm} t \quad (99)$$

$\dot{m}_{adm}$ is calculated with Eqs. 49 or 50 and time t is obtained from Eq. 46.

3.10. Combustion products model

This model allows determining theoretically the composition at equilibrium and combustion product thermodynamic properties of a fuel-air mixture whereas reactive products consist of ten chemical species. The combustion global equation for a ten chemical species system is:

$$\begin{aligned} & m_c C_m H_n O_l + \frac{n + \frac{m}{4} + \frac{l}{2}}{\phi} \left(O_2 + 3.7 N_2 \right) \rightarrow \\ & \rightarrow y_1 H_2O + y_2 H_2 + y_3 OH + y_4 H + y_5 N_2 + y_6 NO + y_7 CO_2 + y_8 CO + y_9 O_2 + y_{10} O \end{aligned} \quad (100)$$

Applying conditions of mass conservation to elements C, H, O and N and reactants we obtain five equations with eleven unknowns (product molar fractions and fuel mass) as shown in following expressions:

$$C:\ y_1 + y_8 = n \tag{101}$$

$$H:\ 2y_1 + 2y_2 + y_3 + y_4 = m \tag{102}$$

$$O:\ y_1 + y_3 + y_6 + 2y_7 + y_8 + 2y_9 + y_{10} = l + 2\frac{\left(n + {}^{m}\!/_{4} - {}^{1}\!/_{2}\right)}{\phi} \tag{103}$$

$$N_2:\ 2y_5 + y_6 = 7.428\frac{\left(n + {}^{m}\!/_{4} - {}^{1}\!/_{2}\right)}{\phi} \tag{104}$$

$$\sum_{i=1}^{10} y_i = 1 \tag{105}$$

Applying the chemical equilibrium to combustion reaction yields six additional algebraic equations to close the system:

$$H_2 \leftrightarrow 2H \qquad K_1 = \frac{y_4^2}{y_2}\left(\frac{p}{p_0}\right) \tag{106}$$

$$O_2 \leftrightarrow 2O \qquad K_2 = \frac{y_4^2}{y_9}\left(\frac{p}{p_0}\right) \tag{107}$$

$$H_2 + O_2 \leftrightarrow 2OH \qquad K_3 = \frac{y_3^2}{y_2 y_9} \tag{108}$$

$$\frac{1}{2}O_2 + \frac{1}{2}N_2 \leftrightarrow 2OH \qquad K_4 = \frac{y_6^2}{y_9^{\frac{1}{2}} y_5^{\frac{1}{2}}} \tag{109}$$

$$H_2 + \frac{1}{2}O_2 \leftrightarrow H_2O \qquad K_5 = \frac{y_1}{y_2 y_9^{\frac{1}{2}}}\left(\frac{p}{p_0}\right)^{-{}^{1}\!/_{2}} \tag{110}$$

$$CO + \frac{1}{2}O_2 \leftrightarrow CO_2 \qquad K_6 = \frac{y_7}{y_8 y_8^{\frac{1}{2}}}\left(\frac{p}{p_0}\right)^{-{}^{1}\!/_{2}} \tag{111}$$

In Eqs. 101 to 111 p is the products pressure, y_1 to y_{10} are the species molar fractions and K1 to K6 are the equilibrium constants which are function of temperature. The system is solved by calculating the partial derivatives with respect to temperature, pressure, and mixture richness through FARG, ECP y PER routines. Applying the first law of thermodynamics to reactants and products depending on the type of combustion process we can find their thermodynamic properties. For constant volume combustion and constant pressure combustion respectively:

$$U_R = U_P \tag{112}$$

$$H_R = H_P \tag{113}$$

3.11. NOx formation model

The nitrous oxides (NOx) produced by a RICE are NO and NO_2 and can be calculated by Zeldovich mechanism. This theory postulates that the production of oxides during the combustion process can be explained through the following chemical reactions:

$$R_1: \; O + N_2 \Leftrightarrow NO + N \tag{114}$$

$$R_2: \; N + O_2 \Leftrightarrow NO + O \tag{115}$$

$$R_3: \; N + OH \Leftrightarrow NO + H \tag{116}$$

A differential equation which allows finding the NO concentration by unit time during combustion, using Zeldovich reactions and considering the possibility of occurrence in both directions can be obtained using the basic theory of chemical kinetics:

$$\frac{d[NO]}{dt} = k_1^+[O][N_2] + k_2^+[N][O_2] + k_3^+[N][OH] - k_1^-[NO][N] - k_2^-[NO][O] - k_3^-[N][OH] \tag{117}$$

Considering the steady state condition for [N], which is equivalent to assume its change is small compared to other interesting species and proceeding in similar mode to the previous case, we obtained:

$$\frac{d[N]}{dt} = k_1^+[O][N_2] - k_2^+[N][O_2] - k_3^+[N][OH] - k_1^-[NO][N] + k_2^-[NO][O] + k_3^-[NO][H] \tag{118}$$

$$\frac{d[N]}{dt} = 0 \tag{119}$$

Substituting Ec. 118 in Ec.117 and using the equilibrium reactions (Ecs. 114 to 116) gets the following expression:

$$\frac{d[NO]}{dt} = \frac{2R_1\left\{1 - \left([NO]/[NO]_e\right)^2\right\}}{1 + \frac{\left([NO]/[NO]_e\right)R_1}{\left(R_2 + R_3\right)}} \tag{120}$$

and Eqs. 114 to 116 are transformed into:

$$R_1 : k_1^+ [O]_e [N_2]_e \quad (121)$$

$$R_2 : k_2^- [NO]_e [O]_e \quad (122)$$

$$R_3 : k_3^- [NO]_e [H]_e \quad (123)$$

Terms enclosed in square brackets represent the chemical equilibrium concentration for corresponding species obtained using routines FARG and ECP. Constants for Zeldovich reactions, depending on temperature, are expressed in the following way.

$$K_1^+ = 7.6E13 \exp(-38000/T) \quad (124)$$

$$K_2^- = 1.5E09 \exp(-19500/T) \quad (125)$$

$$K_3^- = 2.0E14 \exp(-23650/T) \quad (126)$$

Positive sign indicates reaction tendency to form products, while negative sign indicates tendency to form reactants.

3.12. Exhaust gas recirculation (EGR) model

Exhaust gas recirculation (EGR) is a very effective technique employed to reduce nitrogen oxide emissions (Lapuerta, 2000). The method involves replacing a part of the intake air with exhaust gases during the admissions process. In this way a decrease in the amount of air available for combustion reduces the final temperature of combustion, and therefore lowers the production of nitrous oxides emissions. This process can be expressed in the following way:

$$Fuel + Air + \%EGR \longrightarrow Products$$

Recirculated gases will be taken at engine exit and will be introduced into the cylinder at a temperature greater than air admitted. Composition and properties calculation of involved species will be made with routines FARG and ECP.

3.13. Work, mean effective pressure and efficiency

To determine the work produced by closed loop cycle, we consider that it is represented by the loop enclosed area in a p-V diagram. The area was calculated with the trapezoidal rule [42]:

$$A = W = \{p_0 \Delta V + p_1 \Delta V + p_2 \Delta V + p_3 \Delta V .... p_{n-1} \Delta V + p_n \Delta V\} \quad (127)$$

Gas exchange work, indicated work and net work are obtained with the following equations:

$$W_{pum} = W_{exh} - W_{adm} \tag{128}$$

$$W_i = W_{\exp} - W_{com} \tag{129}$$

$$W_{net} = W_i - W_{pum} \tag{130}$$

Net power is obtained through the following expression:

$$\dot{W}_{net} = W_{net} \frac{rpm}{30j} \tag{131}$$

Mean indicated pressure, mean net indicated pressure and mean indicated gas exchange pressure are calculated with equations:

$$mip = W_i / V_d \tag{132}$$

$$mip_{net} = W_{net} / V_d \tag{133}$$

$$mip_{pum} = W_{pum} / V_d \tag{134}$$

Efficiencies are determined with:

$$\eta_i = \frac{W_i}{m_c\ H_i} \tag{135}$$

$$\eta_{i\,net} = \frac{W_{net}}{m_c\ H_i} \tag{136}$$

3.14. MECID computer program

MECID program consist of a routines group for determining: optimum lifting of intake and exhaust valves, gas leakage across piston rings (blow by), composition variation of chemical species that constitute the working fluid, effect of heat transfer, ignition delay, heat release, residual gases and recirculated gases, as well as nitrous oxides formation and pressure drop during intake and exhaust processes due to valve contraction. To calculate all parameters it uses the mathematical relationships and models proposed in preceding paragraphs. Figure 2 shows the MECID program flow diagram.

Using the ENGINE routine the program requested information on engine works site conditions (temperature and pressure) and on engine geometric characteristics (piston bore and stroke; valves diameter, maximum lift and average discharge coefficient, compression ratio, connecting rod length to crank radius ratio and engine speed) With VALVE routine requested information on calculation conditions (calculation angular interval, advance and delay of intake and exhaust valve closure, richness, residual gases ratio, recirculated gas percentage and cylinder wall average temperature). Selected heat transfer, ignition delay and heat release models which will be used in the application, as well as the angles when combustion begins and ends, the program allows to perform studies to determine what should be the working conditions to ensure maximum fuel energy utilization the influence of different variables on engine operation. Results of more important studies will be presented in next paragraph.

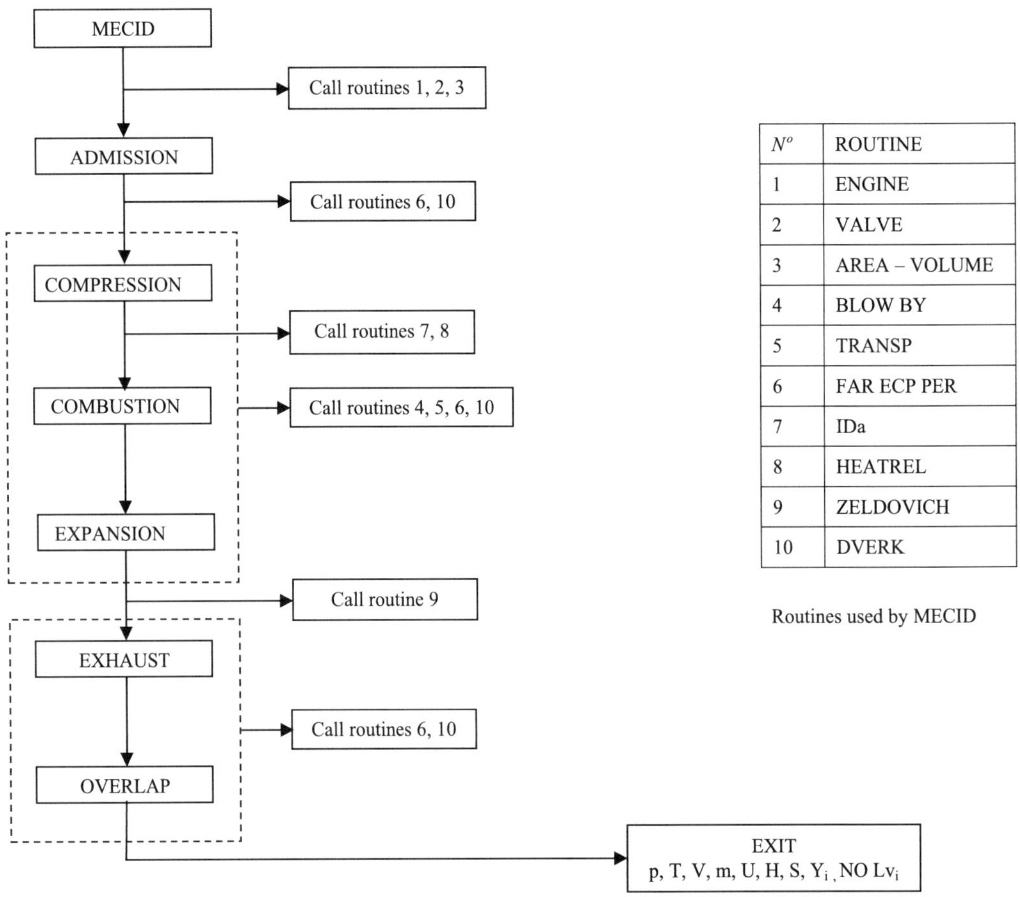

N°	ROUTINE
1	ENGINE
2	VALVE
3	AREA – VOLUME
4	BLOW BY
5	TRANSP
6	FAR ECP PER
7	IDa
8	HEATREL
9	ZELDOVICH
10	DVERK

Routines used by MECID

Figure 2. Block diagram and routines used by MECID computer program

4. Results and discussion

4.1. Effect of burning process beginning on CIE operating parameters

For combustion duration constant and varying combustion start we obtained the p and T vs. ϕ diagrams shown in Figure 3. We note that if the combustion process is advanced too much, from 360° to 310°, it reached a high maximum pressure while the piston is at some point in the compression stroke. This brings as a consequence, greater work consumption and reduced cycle efficiency. If combustion beginning delays too much, 360° to 410°, the maximum pressure is very low. If combustion beginning advance is such that the process occurs around TDC we obtain intermediate maximum pressures during expansion process and a higher maximum temperature, representing greater thermal energy available to do work. In view of the above, for subsequent studies we will take a 345° angle as combustion beginning angle. Maximum temperature is obtained for a greater angle than the maximum pressure one because the burning model uses constant combustion duration.

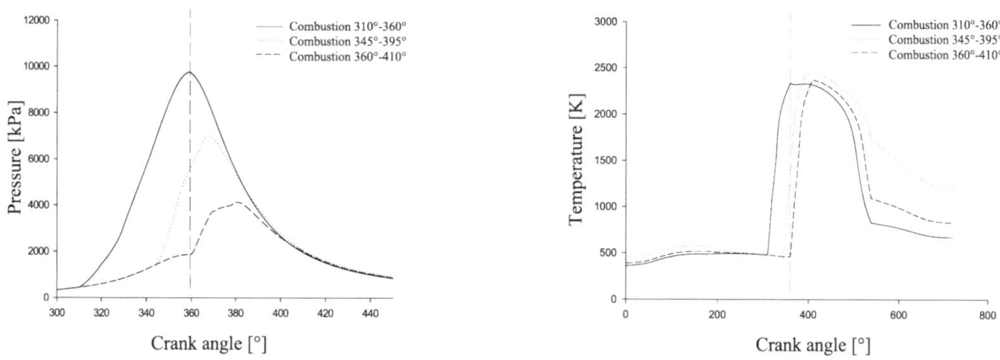

Figure 3. p vs ϕ and T vs ϕ diagrams for various combustion beginning angles

Figure 4 shows how varies the accumulated burned mass fraction and the instant burned mass fraction depending on combustion start angle. Burned mass is responsible for pressure and temperature rapid increase during combustion process. It is observed that peak values are the same for all three cases considered. This is because richness and volumetric efficiency remain unchanged.

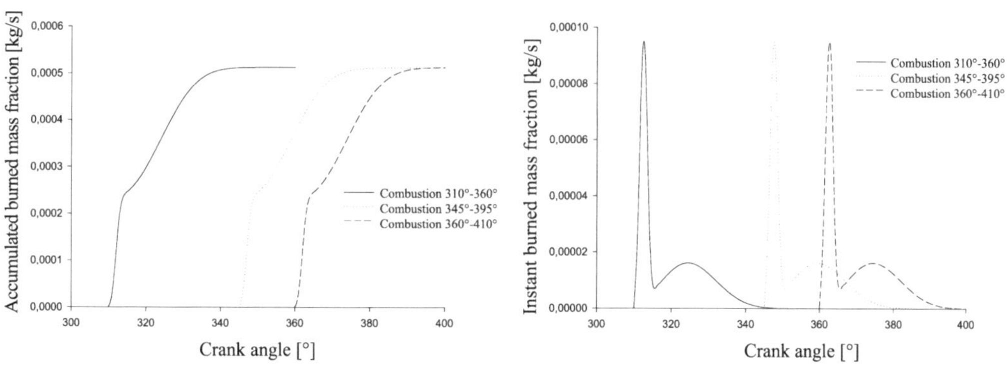

Figure 4. Accumulated and instant burned mass fraction for various combustion beginning angles

4.2. Effect of combustion process duration on CIE operating parameters

In present case, the start of combustion process is kept constant. Figure 4 shows that at lower combustion duration the greater the maximum pressure reached since a great amount of heat is released in a short time period, as seen by the burned curves slope. In these curves can be noted that in considered cases, the maximum burned fractions do not change because the richness was kept constant and the obtained volumetric efficiency remained almost constant, therefore the total in-cylinder mixture mass is the same for all referred cases.

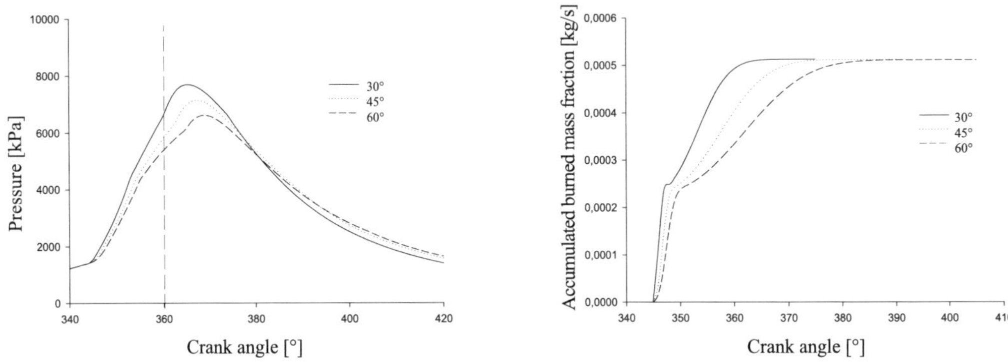

Figure 5. Pressure and accumulated burned fraction for different combustion duration

Table 4 presents the principal results obtained for operation parameters. From Table 4 it can be concluded that the more efficient cycle producing higher power, is one in which combustion takes 45° and therefore this duration will be used in future studies.

Combustion process duration	30º	45º	60º
η_v	0.72	0.72	0.72
W_i [kJ]	0.60	0.73	0.75
W_{net} [kJ]	0.57	0.65	0.64
$\dot{W}_{net}$ [kW]	9.56	10.98	10.72
mip [kPa]	994.75	1200.21	1232.88
$\eta_{i\ net}$	0.35	0.40	0.39
P_{max} [kPa]	7689.17	7134.93	6614.27
T_{max} [K]	2235.64	2400.71	2496.92

Table 4. Summary of main results obtained when varying combustion process duration.

4.3. Effect of angular velocity on CIE operating parameters

Pressure variation depending on crankshaft rotation angle when engine speed change, is shown in Figure 6. It can be noted from this figure that higher pressures are reached when the engine operates at lower speeds. This is because while the engine works at lower speeds, greater will be its volumetric efficiency. If the richness is kept constant, the greater the mass which is burned during combustion process and thus more energy will be released.

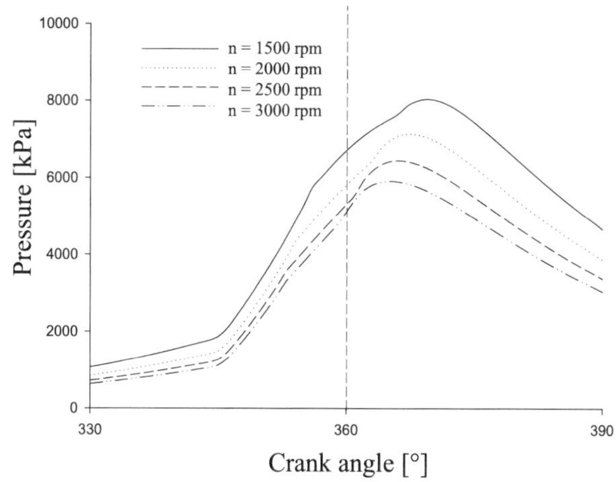

Figure 6. In-cylinder pressure for various engine speeds

Engine speed	1500 rpm	2000 rpm	2500 rpm
η_v	0.92	0.69	0.55
β	50.57	45.33	40.89
W_i [kJ]	1.42	1.18	1.03
$\dot{W}_{net}$ [kW]	14.63	16.01	17.28
mip [kPa]	2314.47	1922.25	1678.68
η_i	0.71	0.74	0.77
p_{max} [kPa]	6703.23	6034.92	5506.50

Table 5. Summary of main results obtained varying engine speed.

Table 5 presents the principal results obtained for operating parameters. From this table it can be concluded that at higher rpm volumetric efficiency is reduced due to increased air velocity in the intake system and increased pressure friction losses. Additionally, as there is less time to fill the cylinder, which results in lower air intake, the maximum pressure reached is reduced as well as the indicated work. Net power increases due to the influence of increasing rpm. Indicated efficiency tends to increase with increasing rpm, because burning the same fuel amount at higher rpm and net power is higher. Figure 6 shows that maximum pressure is obtained in all cases close to the TDC. One can also appreciate that the higher the rpm, the mass fraction burned during the premixed phase and the mean indicated pressure are reduced

4.4. Effect of compression ratio on CIE operating parameters

Now we will consider the effect of varying the compression ratio on the main engine operating parameters assuming valve inlet pressure equal to atmospheric pressure. Figure 7 shows the in-cylinder pressure variation and the accumulated burned mass fraction for various compression ratios and in Table 6 we can see the summary of main results obtained for different compression ratios.

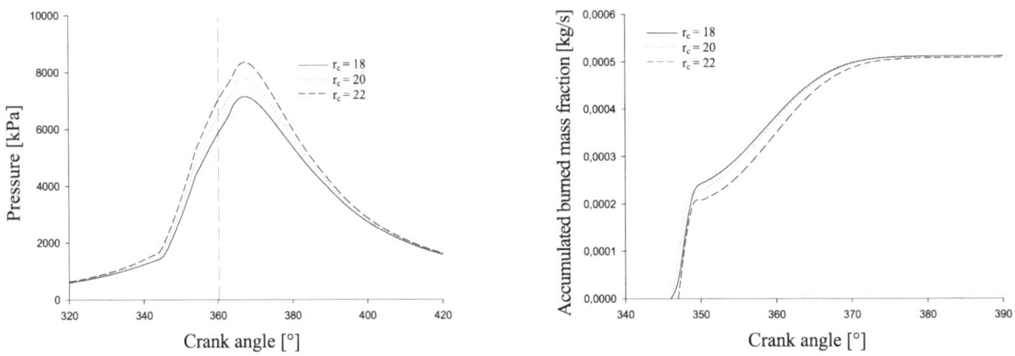

Figure 7. Pressure and accumulated burned fraction for different compression ratios

By analyzing Figure 7 and Table 6 we can conclude that: i) volumetric efficiency is unaffected by compression ratio since is not dependent on it, ii) at higher compression ratios there is a reduction of burned mass fraction in the premixed phase, iii) there is an indicated work increase because maximum pressure and temperature increase, resulting in increased efficiency and mean pressures, iv) pumping work increases when compression ratio increases as pressures are greatest during exhaust, v) cycle net work increases because net work and power increase at higher rates than pumping work.

Compression ratio	18	20	22
η_v	0.72	0.72	0.72
β	45.33	42.68	40.24
W_i [kJ]	0.73	0.75	0.77
$\dot{W}_{net}$ [kW]	10.99	11.36	11.68
mip [kPa]	1200.22	1233.52	1258.36
η_i	0.45	0.46	0.47
P_{max} [kPa]	7134.93	7791.11	8349.81
T_{max} [K]	2400.71	2399.34	2392.72

Table 6. Summary of main results obtained varying compression ratio.

4.5. Effect of atmospheric pressure on CIE operating parameters

It is well known that increasing the engine speed the start of combustion must anticipate in order that maximum pressures remain nearly constant and occur near TDC. For this reason in present study the combustion duration will remain constant and, its beginning will be advanced. Figure 8 shows the indicator diagram for various engine speeds. In Figure 8 which shows the indicator diagram for various engine speeds, it is observed that the maximum pressure in all studied cases occurs near TDC and its value is approximately constant. Since optimum injection angle for each speed is not known it is not easy to superimpose the curves which is what is desired in an engine in order to obtain maximum power and efficiency in a wide speeds range.

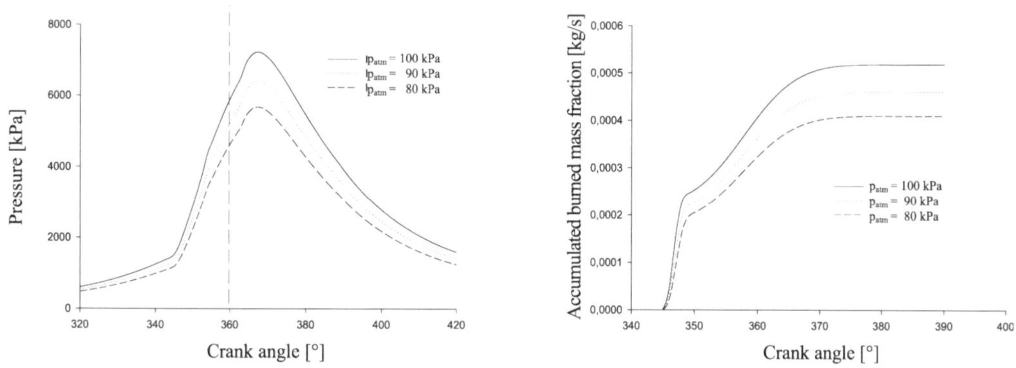

Figure 8. Pressure and accumulated burned fraction for various atmospheric pressures

Atmospheric pressure	100 kPa	90 kPa	80 kPa
η_v	0.72	0.72	0.72
β	45.25	45.93	46.59
W_i [kJ]	0.74	0.65	0.58
$\dot{W}_{net}$ [kW]	11.13	9.90	8.80
mip [kPa]	1217.03	1073.99	948.11
η_i	0.45	0.44	0.44
p_{max} [kPa]	7231.67	6406.81	5669.70
T_{max} [K]	2401.91	2391.06	2380.09

Table 7. Summary of main results obtained varying engine speed.

The summary of results in Table 7 shows that increasing the engine speed i) volumetric efficiency is reduced which is the expected behavior for CIE, ii) mass fraction burned during the premixed phase is reduced and iii) mean pressure decreases possibly due to a combustion beginning advance not enough for the increase in speed considered.

5. Conclusions

A methodology to theoretically determine the thermodynamic working cycle of direct injection compression ignition engines was presented.

A computer program based on the application of the above methodology was introduced.

Results of several studies performed with the program was presented and discussed. From some results we can conclude:

- As the engine speed increases the start of combustion should anticipate maintaining nearly constant maximum cycle pressure as well as the angle at which this occurs.
- At higher compression ratios, the greater the potency and mean effective pressure, this coincides with the theoretical internal combustion engines behavior.
- At higher atmospheric pressure the engine volumetric efficiency will increase which ensures a greater air intake and thus an increase in produced power.

Notation

A	area
c	piston stroke
Cd	discharge coefficient
Cp	constant pressure specific heat
Cv	constant volume specific heat
d( )	preceding term ordinary derivative
D_p	piston diameter
d_v	valve diameter
E	energy
e	specific energy
f	fuel air ratio
h	specific enthalpy
H	enthalpy
h_c	convection heat transfer coefficient
h_c	mean convection heat transfer coefficient
Hi	lower heating value
ID	ignition delay
j	engine strokes
k	conduction heat transfer coefficient
L_v	valve lift
m	mass
$\dot{m}$	mass flow
$\tilde{m}$	adimensional mass
mip	mean indicated pressure
Nu	Nusselt number
p	pressure
$\tilde{p}$	adimensional pressure
p_{down}	downstream pressure
p_{up}	upstream pressure
Q	heat
$\tilde{Q}$	adimensional heat
$\dot{Q}$	heat flow
q	heat per unit mass

R	gas constant
r_c	compression ratio
Re	Reynolds number
RLA	connecting rod length to crank radius ratio
rpm	engine speed rpm
T	temperature
$\tilde{T}$	adimensional temperature
t	time
u	specific internal energy
U	internal energy
V	volume
$\tilde{V}$	adimensional volume
v	specific volume
V_{mp}	piston mean velocity
W	work
$\dot{W}$	power
w	specific work
X_b	burned mass fraction
y_i	i gas fraction

Greek letters

β	premixed phase burning factor
γ	polytropic coefficient
$\partial(\,)$	preceding term partial derivative
η	efficiency
μ	viscosity
π	3.1416
ϱ	density
ϕ	crankshaft rotation angle
ω	angular velocity
φ	equivalence ratio (mixture richness)

Subindex

a	air
adm	admission
ar	real air
at	theoric air
bb	blow by
cc	combustion chamber
cil	cylinder
com	combustion
com	compression
d	displaced

dif	diffusive phase
e	input
EGR	exhaust gas recirculated
exh	exhaust
exp	expansion
f	fuel
gr	residual gasses
i	indicated
lost	lost
max	maximum
net	net
no comb	without combustion
P	products
pre	premixed phase
pum	pumping
R	reactants
ref	reference condition
s	output
sto	stoichiometric
Tot	total
v	volumetric
vc	control volume
wall	cylinder wall

Author details

Simón Fygueroa
National University of Colombia, Colombia,
Turin Polytechnic Institute, Italy,
Polytechnic University of Valencia, Spain,
Mechanical Engineering Department, University of the Andes, Mérida, Venezuela,
Pamplona University, Mechanical Engineering Department, Pamplona, Colombia

Carlos Villamar
University of the Andes, Mérida, Venezuela,

Olga Fygueroa
University of the Andes, Mérida, Venezuela,
Polytechnic University of Valencia, Spain
Altran-Nissan Technical Center, Barcelona, Spain

6. References

[1] Jovaj MS. Automotive vehicle engines. Moscu: Editorial MIR; 1982.

[2] Payri F. and Desantes JM. Motores de combustión interna alternativos. Valencia, (Spain): Servicio de Publicaciones Universidad Politécnica de Valencia; 2011.

[3] Rajput RK. Internal combustion engines. New Delhi: Laxmi Publications Ltd; 2007.
[4] Holt DJ. The diesel engine. Warrendale (PA): Society of Automotive Engineers; 2004.
[5] Ganesan V. Internal combustion engines. New Delhi: Tata McGraw-Hill Publishing Company Ltd; 2008.
[6] Arcoumanis C, Bicen AF, Whitelaw JH. Effect of inlet parameters on the flow characteristics in a four-stroke model engine. Warrendale, (PA): Society of Automotive Engineers, 1982.
[7] Benson R. The thermodynamics and gas dinamics of internal combustión engines. Oxford, (England): Clarendon Press; 1986.
[8] Depcik C, Assanis D. A universal heat transfer correlation for intake and exhaust flows in a spark ignition internal combustion engine. SAE Transaction 2002-01-0372, 1972.
[9] Woschni G. Computer programs to determine the relationship, and thermal load in diesel engines. SAE Technical Paper 650450, 1965.
[10] Annand WJ, Pinfold D. Heat transfer in the cylinder of a motored reciprocating engine. SAE Technical Paper 800457, 1980.
[11] Sitkei G. Heat Transfer and termal loading in internal combustión engine. Budapest: Akadémiai Kiadó; 1974.
[12] Gosman AD. Computer modeling of flow and heat transfer in engines, progress and prospects. In Proceedings of COMODIA 1985; Tokio; 1985.
[13] Meisner S, Sorenson S. Computer simulation of intake and exhaust manifold flow and heat transfer. SAE Technical Paper 860242, 1986.
[14] Taylor CF. The internal-combustion engines in theory and practice: combustion, fuels, materials, design. Cambridge, (MA): MIT Press; 1982.
[15] Heywood JB. Internal combustion engine fundamentals. New York: McGraw-Hill; 1988.
[16] Assanis DN, Filipi ZS, Fiveland SB, Syrimis M.A Predictive ignition delay correlation under steady-state and transient operation of a direct injection diesel engine. Eng Gas Turbines Power 2003; 125(2).
[17] Liu H. Simulation model for steady state and transient cold starting operation of diesel engines. ETD Collection for Wayne State University. Paper AAI3037106, Michigan, 2001.
[18] Woschni G. A universally applicable equation for the instantaneous heat transfer coefficient in the internal combustion engine. SAE Technical Paper 670931, 1967.
[19] Hardenberg HO, Hase F. An empirical formula for computing the pressure rise delay of a fuel from its cetane number and from the relevant parameters of direct-injection diesel engines. SAE Technical Paper 790493, 1979.
[20] Gardner T, Henein N. Diesel starting: a mathematical model. SAE Technical Paper 880426, 1988.
[21] Way RJ. Programs for determination of composition and thermodynamic properties of combustion products for internal combustion engines calculations. Proceedings of the Inst Mech Engrs 1977; 190(60/76): 687-697.
[22] Desantes JM, Lapuerta M. Fundamentos de combustión. Valencia, (Spain): Servicio de Publicaciones Universidad Politécnica de Valencia; 1991.
[23] El-Mahallawy F, El-Din HS. Fundamentals and technology of combustion. London: Elsevier Science Ltd; 2002.

[24] Jarquin G, Polupan G, Rodríguez GJ. Cálculo de los productos de combustión empleando métodos numéricos. Mecánica Computacional 2003; 22: 2442-2452.

[25] McBride BJ, Gordon S. Fortran IV program for calculation of thermodynamic data. NASA Publication TN-D-4097, 1967.

[26] Gordon S, McBride BJ. Computer program for calculation of complex chemical equilibrium compositions, rocket performance, incident and reflected shocks, and Chapman-Jouguet detonations. NASA Publication SP 273, 1971.

[27] Harker JH. The Calculation of equilibrium flame gas composition. Journal Inst Fuel 1967; 40: 206-210.

[28] Harker JH, Allen DA. The calculation of the temperature and composition of flame gases. Journal Inst Fuel 1969; 42: 183-187.

[29] Olikara C, Borman GL. A computer program for calculating properties of equilibrium combustion products with some applications to IC. engines. SAE Technical Paper Nº 750468, 1975.

[30] Agrawal DD, Sharma SP, Gupta CP. The calculation of temperature and pressure of flame gases following constant volume combustion. Journal Inst Fuel 1977; 50: 121-124.

[31] Agrawal D, Gupta CP. Computer program for constant pressure or constant volume combustion calculations in hydrocarbon-air ssystems. Transactions of the ASME 1977: 246-254.

[32] Lapuerta M, Armas O, Hernández J. Diagnosis of DI diesel combustión from in-cylinder pressure signal by estimation of mean thernodynamic properties of the gas. Applied Thermal Engineering, 1997.

[33] Araque J, Fygueroa S, Martín M. Modelado de la combustión en un MECH-CFR. In Memorias del IV Congreso Nacional de Ingeniería Mecánica, Mérida (Venezuela), 2001.

[34] Ferguson C, Kirkpatrick A. Internal combustion: applied thermosciences. New York: John Wiley and Son; 2001.

[35] Benson R, Annand W, Baruah P. A simulation model including intake and exhaust systems for a single cylinder four stroke cycle spark ignition engine. International Journal of Mechanical Sciences 1975; 17(2): 97-124.

[36] Woodward J. Air standard modelling for closed cycle diesel engines. Proc Instn Mech Engrs 1995; 209.

[37] Hull TE, Enright WH, Jackson KR. Runge-Kutta research at Toronto. Journal Applied Numerical Mathematics 1996; 22: 225-236.

[38] Assanis DN. Valve event optimization in a spark-ignition engine. J Eng Gas Turbines Power 1990; 112(3): 341-348.

[39] Watson N, Pilley AD, Marzouk M. A Combustion correlation for diesel engine simulation. SAE Technical Paper 800029, 1980.

[40] Miyamoto T, Hayashi K, Harada A, Sasaki S, Akagawa H, Tsujimura K. Numerical simulation of premixed lean diesel combustion in a DI engine. In Proceedings of COMODIA 98, Kioto, 1998.

[41] Annand WJ. Heat transfer in the cylinders of reciprocating internal ccombustion engines. Proceedings of the Institution of Mechanical Engineers 1963; 177(1): 973-996.

[42] Chapra S, Canale R. Métodos numéricos para ingenieros. México: Mc Graw Hil; 2002.

Chapter 5

The Effect of Injection Timing on the Environmental Performances of the Engine Fueled by LPG in the Liquid Phase

Artur Jaworski, Hubert Kuszewski, Kazimierz Lejda and Adam Ustrzycki

Additional information is available at the end of the chapter

http://dx.doi.org/10.5772/54604

1. Introduction

Gaseous fuels are widely used in internal combustion engines because of their properties and benefits. This is mainly due to their smaller burden on the environment and lower prices. They are used not only to power the traction engine in passenger cars or buses, but also in other applications, for example, to power the engines driven electric generators. Ever-expanding chain of filling stations increases the availability of these fuels, which also affects the development of such fuelling systems. Because of their advantages fuel injection systems with their own drivers are used increasingly. It provides a non-collision work of mentioned drivers with a gasoline engine drivers of the vehicles, providing the performance of an engine running on gas fuel comparable to one running on gasoline.

The use of LPG injection in the liquid phase makes it possible to have more precise fuel delivery, and even more limits amount of pollution emitted by the engine to the environment. The fuelling for internal combustion engines with LPG in liquid phase using injection system into the intake manifold is a solution very similar to conventional fuel systems (Hyun et al., 2002; Lee et al., 2003). This kind of fuelling to the cylinders allows to obtain performances of engine comparable to the performances obtained using petrol and diesel fuel system.

2. Adaptation of engine to fuelling with lpg in liquid phase at sequence injection

The gas fuel system allowed the gas injection in the liquid phase was adapted to the engine, which in the original version was a compression ignition engine with symbol MD-111. This engine is a 6-cylinder Diesel engine with direct injection. The combustion chamber is made in the piston bowl and has a toroidal shape. By reducing the compression ratio and ignition system implementation, in the first stage, the engine to gas fuelling in the volatile phase has been adapted. The combustion chamber has been redesigned, gaining "cup" shape, whereby the compression ratio was also reduced from 16.5 to 9. The cylinder head was changed enabling to implementation a spark plug. To control the engine load, in the inlet system the throttle valve was installed.

The construction and operational parameters of modernized engine are summarized in Table 1. Next, the engine was further modernized in order to adapt gas fuel system enabling to sequence injection in the liquid phase.

As a gas system, the Vialle system was used (Fig. 1). The installation consists of:

- Electronic Control Unit (ECU) of LPG fuel system,
- tank with fuel pump,
- LPG injectors,
- fuel pipes.

The system was developed for mating with the ECU of petrol engine in the system MASTER-SLAVE. In this system, ECU of LPG fuel system uses injection duration determined by ECU of petrol engine for calculating the opening duration of gas injectors. Since the MD-111E engine did not have the electronic control unit, the primary issue was to develop a control unit, generating suitable values of the injection duration for the ECU of LPG fuel system.

The main components of a LPG, i.e. propane and butane, have low boiling points. These temperatures are respectively 231 K and 272.5 K, and are lower than the average ambient temperatures encountered during engine operation. Especially high temperatures are in the engine compartment of the vehicle (hot zone), where temperatures reach about 350 K. This causes the temperature rise in the fuel system, which leads to evaporation of fuel in the fuel pipes and formation of vapour-locks (Cipollone & Villante, 2000, 2001; Dutczak et al., 2003). Keeping gas in the liquid phase in such difficult conditions requires its compression. To obtain a stable injection of LPG in the liquid phase, the system was equipped with a pressure monitoring system. This function is performed by the pump (Fig. 2) placed in the fuel tank (Fig. 3) and pressure regulator (Fig. 4). The pressure regulator maintains the pressure in the supply system higher than the pressure in the tank. It allows to delivery fuel to injectors in liquid phase at every conditions of engine operation.

The liquid gas is pumped through a suitably shaped diaphragm pump. The pump has 5 chambers, which are integrated with the suction valves and power valve. The pump motor is a brushless alternating-current motor with permanent magnets. It is powered by DC, which is transformed into AC with a frequency controlled by an electronic control unit located in the assembly lid. The motor can be rotated with five different speeds 500, 1000, 1500, 2000, 2800 rpm. Speed control is realized by the ECU of LPG fuel system, depending on engine speed and load (injection duration). The pressure regulator is located between tank and gas injectors.

No.	Name	Value
1	Producer	WS Mielec
2	Trademark	MD-111E
3	Company name	„PZL Mielec"
4	Type	MD-111E
5	Work cycle	4-stroke
6	Cylinder diameter	127 mm
7	Piston stroke	146 mm
8	Engine capacity	11097 cm^3
9	Number and basic engine design	6-cylinder, in-line
10	Firing order	1-5-3-6-2-4
11	Type of combustion system	spark ignition
12	LPG fuel system	mixer, electronic control of injection process
13	Compression ratio	9 : 1
14	Minimum cross section area: - inlet port - exhaust port	 1250 mm^2 950 mm^2
15	Cooling system	liquid
16	Type of cooling liquid	Ethylene/Propylene Glycol Heat-Transfer Fluid
17	Cooling pump	impeller
18	Radiator and fan	pipe cooler, downcast ventilator with viscose clutch (EATON type)
19	Maximum outlet temperature at radiator	95°C
20	Inlet and fuel system	maximum limit of negative pressure inside inlet manifold at reference point 4.5 kPa
21	Supercharging system	not installed
22	Mixer fuel system: reducer-evaporator mixer type gas dosing system	 Tartarini GP-150 MS-1 WS-Mielec electronic actuator in closed system with oxygen sensor
23	Cold start unit	Direct starting with electrical starter motor powered by battery, power 4.4 kW, voltage 24 V, type R 22
24	Maximum lift of valves	13,3 mm

No.	Name	Value
25	Inlet valve timing: - opening - closing Exhaust valve timing: - opening - closing	 8 deg. before TDC 52 deg. after BDC 46 deg. before BDC 20 deg. after TDC

Table 1. Technical and operational data of MD-111E engine

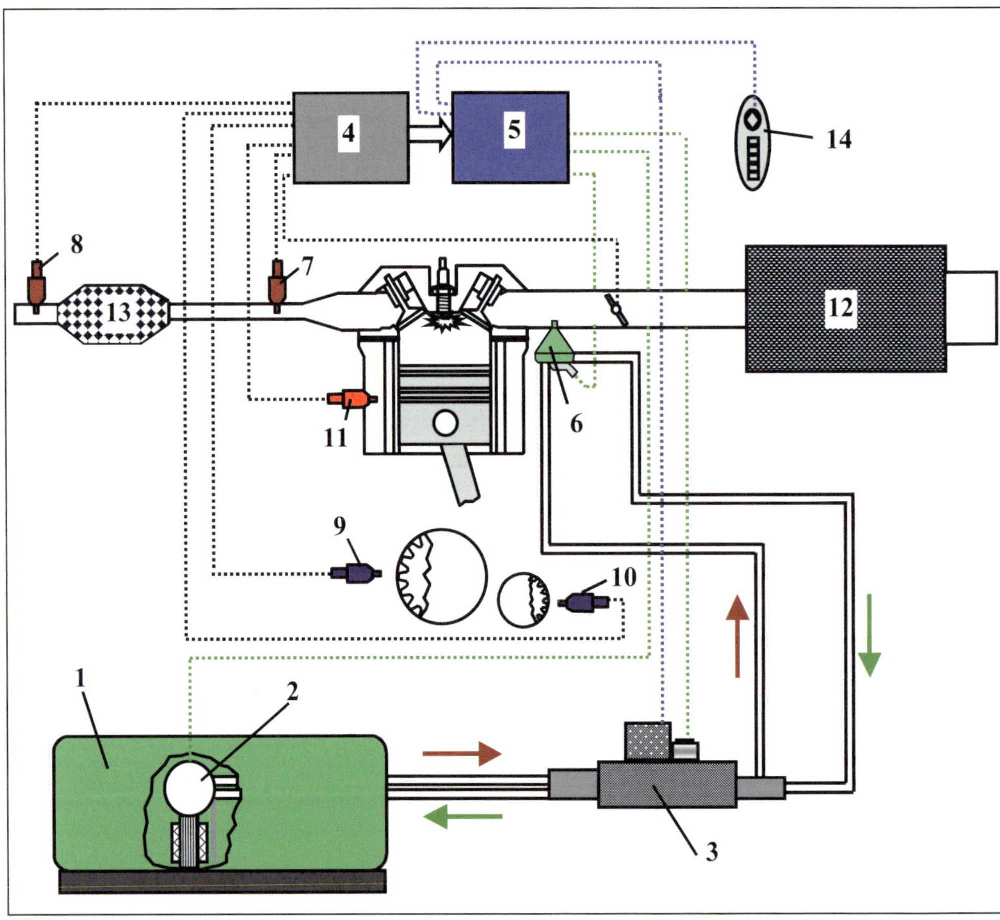

Figure 1. Scheme of VIALLE fuelling system (Vialle, 2001): 1 – LPG tank, 2 – LPG pump, 3 – fuel pressure regulator, 4 – petrol ECU, 5 – LPG ECU, 6 – LPG injector, 7 – first oxygen sensor, 8 – second oxygen sensor, 9 – engine speed sensor, 10 – camshaft position sensor, 11 – coolant temperature sensor, 12 – air filter, 13 – exhaust gas catalyst, 14 – fuel type switch

Figure 2. Unit of LPG pump (Vialle, 2001)

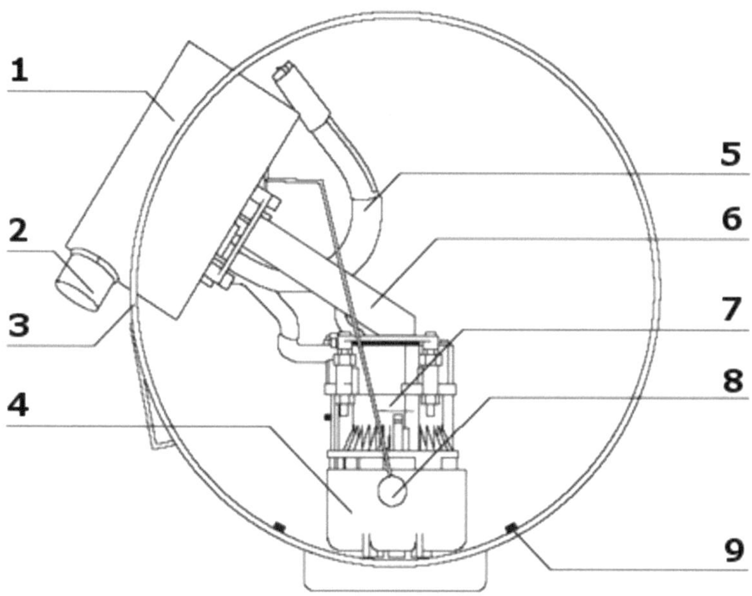

Figure 3. Scheme of tank with fuel pump (Vialle, 2001): 1 – pump body, 2 – inlet of body, 3 – tank wall, 4 – distance sleeve, 5 – inlet pipe, 6 – pump holder, 7 – pump, 8 – float, 9 – magnet

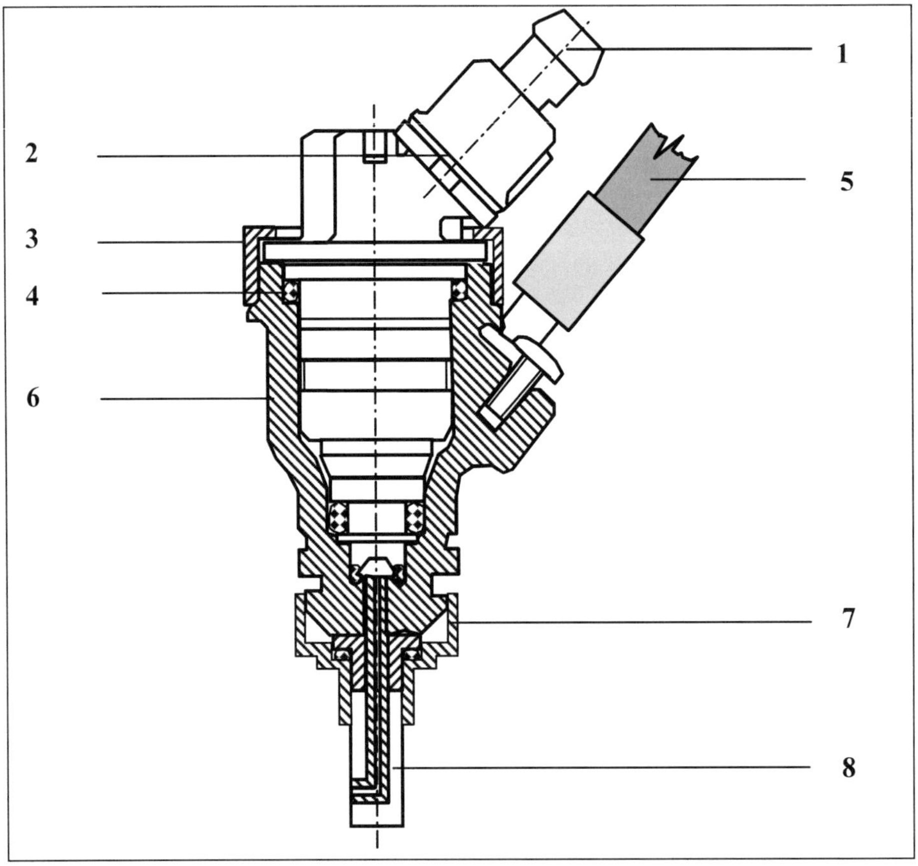

Figure 4. Scheme of LPG liquid phase injector (Vialle, 2001): 1 –electrical connection, 2 – injector body, 3 – ring fixing the injector to body, 4 – o-ring, 5 – fuel inlet socket, 6 – injector housing, 7 – adapter, 8 – outlet pipe

The pressure regulator includes solenoid valve opened and closed when the output valve of the tank is turned on. The pressure regulator is also a pressure control module and the pressure sensor. The liquid gas flows through the valve to the injectors and the excess returns via a pressure regulator to the tank. The pressure is keeping by the controller by 5 bars higher than the pressure in the tank and can be 7-30 bars (Cipollone & Villante, 2000; Vialle, 2001). For the injection of gas in liquid phase there were used the low-pass injectors which distinct from top-pass injectors commonly used in petrol fuel systems, the fuel is delivered below the injector coil. This causes less heating of the gas from the coil, which favors keeping the liquid phase in the injector. To prevent coarse pollutions the filter was placed before the inlet of gas into the injector (Fig. 4). Due to the low resistance of the injector coil, the pulse control was applied to reduce the currents flowing during operation of the injector. The gas is fed to the injectors with synthetic pipes which are fixed with reinforced plates to each other and are locked with screws. Because the exhaust manifold is located above the intake manifold, the injectors were mounted into the intake manifold from below near to the cylinder head. It allows to lead the gas almost directly onto the inlet valve.

3. Test stand and research method

The goal of realized study was to determine the effect of injection timing on ecological parameters of the engine. The test stand with dynamometer has been equipped with the following functional units and measurement systems:

- hydraulic dynamometer Schenck D 630 with the control system,
- exhaust gas analysis system consisting of the following elements:
  - Chemiluminescent NOx analyzer of the type Pierburg CLD PM 2000,
  - Flame ionization hydrocarbon analyzer FID HC type of PIERBURG PM 2000,
  - four-gas exhaust gas analyzer (CO, CO_2, HC, O_2) of the type Bosch BEA 350 equipped with function for calculating the ratio of actual AFR to stoichiometry (Lambda) for the various fuels,
- LPG fuel consumption measuring system with Coriolis sensors,
- flow measurement system of the intake air,
- temperature measurement system,
- pressure measurement system.

The engine mounted to the test stand (Fig. 5) is shown in figure 6.

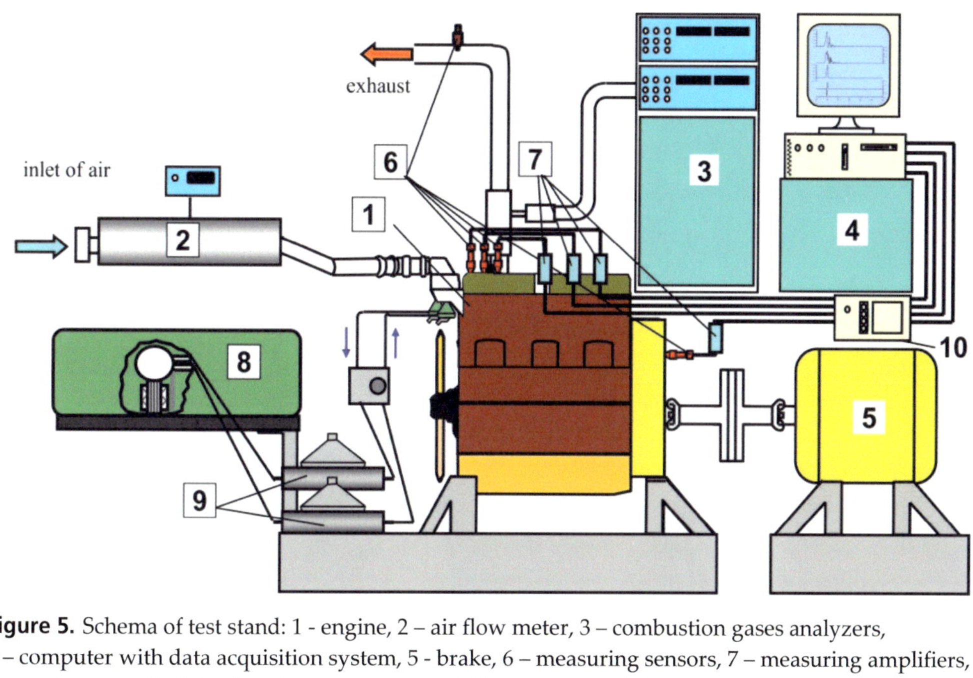

Figure 5. Schema of test stand: 1 - engine, 2 – air flow meter, 3 – combustion gases analyzers, 4 – computer with data acquisition system, 5 - brake, 6 – measuring sensors, 7 – measuring amplifiers, 8 – container of LPG, 9 – fuel flow measurement, 10 – separator of signal

During tests the measurements of following exhaust ingredients were made: oxides of nitrogen (NOx), hydrocarbons (HC) and carbon monoxide (CO). Additionally, engine noise level was determined. Primarily the measurements was conducted for engine speed of n – 1500 rpm, required for co-operation with a power generator and with different loads.

An important parameter that affects both the operating parameters and the exhaust toxicity for sequential injection is the start of fuel injection (Hyun et al., 2002; Oh et al., 2002). For this reason a large part of the measurements was the analysis of the impact of the start of injection on obtained engine parameters and the emission of toxic ingredients in exhaust gases. The tests were performed with single and double injection. The start of injection was changed within the range shown in figure 7 and 8.

Figure 6. MDE-11 engine with sequence LPG injection system (for liquid phase) during test on stand

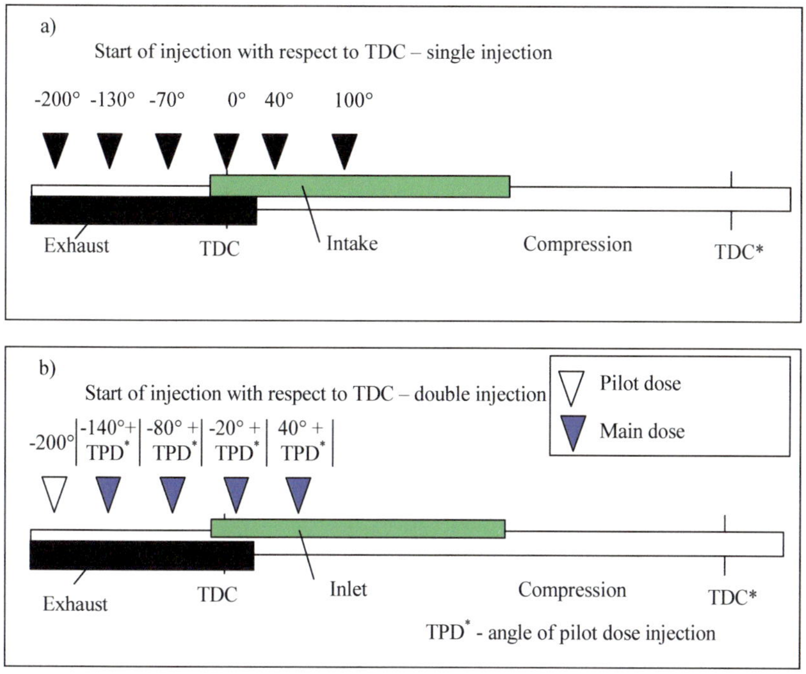

Figure 7. Tested injection starts with first signal disk sensor position: a) for single injection, b) for dual injection

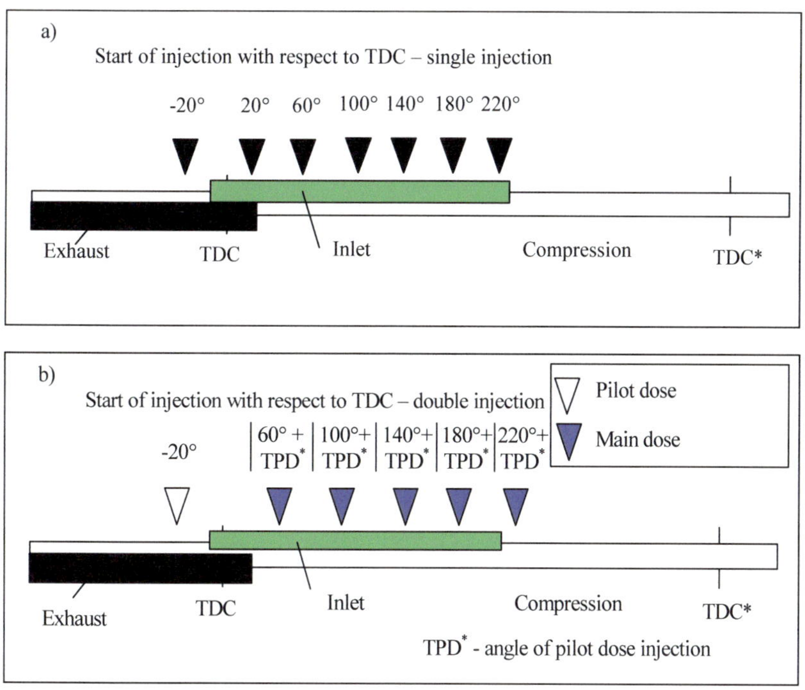

Figure 8. Tested injection starts with second signal disk sensor position: a) for single injection, b) for dual injection

The measurement of noise level was realized with AS-120 meter located at a height of 1 m and 1 m from the engine on the side of electrical starter motor. The sound level was measured using a filter correction L_A [dB] and without correction L [dB]. The microphone for sound recording was placed in the axis of the engine, between 3 and 4 cylinder, at distance of 1 m from the valve cover. There was used microphone AKG C1000S (Shure Beta-58) for sound recording cooperating with amplifier Behringer MX1804X and octave filter RFT OF 101-01000. The recording was performed with 16-bit sound card.

4. Test results

Fig. 9 and 10 presents the performance of the engine, and fig. 11-14 shows the contour map (generalized performance map) for specific fuel consumption and the concentration of carbon monoxides, oxides of nitrogen and hydrocarbons. As we can see the maximum brake torque of the engine is larger than 770 Nm at an engine speed of 900 rpm and the maximum brake power of the engine is 125 kW at an engine speed of 1700 rpm.

Specific fuel consumption for an engine speed of 1500 rpm which is relevant to power generator is lowest at the maximum load and amounts to approximately 265 g/kWh. For this engine speed, the maximum CO concentration is approximately 0.3% and is higher at large loads. The concentration of NOx for mentioned engine speed ranges is from 40-550 ppm, reaching 160 -250 ppm for the large and medium loads. Hydrocarbon concentration amounts to from 15 ppm at small loads, up to 75 ppm at loads close to maximum.

The relationship between the injection starts of the LPG in liquid phase into a inlet manifold pipes and the concentration of CO_2, CO, HC and NOx in the exhaust is shown on figures 15 and 16. At the starts of injection carried out at the opening of the intake valve, an increase in the concentration of NOx and hydrocarbons HC was observed versus the injection starts realized before opening the intake valve (fig. 15). The concentrations of CO and CO_2 were undergone a slight changes in this case. At the injection starts carried out at the opening of the inlet valve is visible increase in the concentration of NOx at injection starts carried out from about 60 to 100 CA deg. after TDC during the intake stroke. Moreover an increase in the concentration of hydrocarbon HC and carbon monoxide CO at the injection starts carried out in the phase of closing the inlet valve from about 140 to 180 CA deg. after TDC at inlet stroke was observed (fig. 16). The injection realized at closing the inlet valve is also connected with the reduction of CO_2 concentration.

Basing on the results for engine parameters and concentrations of hydrocarbons HC, oxides of nitrogen NOx, carbon monoxide CO in exhaust gas a right specific emissions were calculated. The calculation course of the specific emission was determined based on a set of International Standards ISO 8178 (ISO, 1999-2001). The calculation results are shown in figures 17-22.

The injection starts realized at closing the inlet valve (fig. 17 and 20) cause the increase in specific hydrocarbons emissions. Specific hydrocarbons emission decreases with increasing the injection duration (fuel quantity). Moreover we can see that specific NO_X emission increases with long injection durations (higher load) and the injection starts realized at the opening of the inlet valve (fig. 18 and 21).

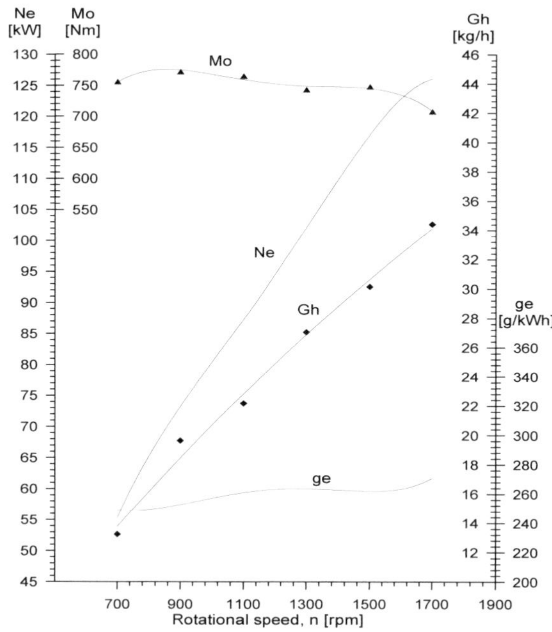

Figure 9. MD-111E engine WOT diagram for double injection: Ne – engine power, Mo – torque, Ge – fuel consumption, ge – specific fuel consumption

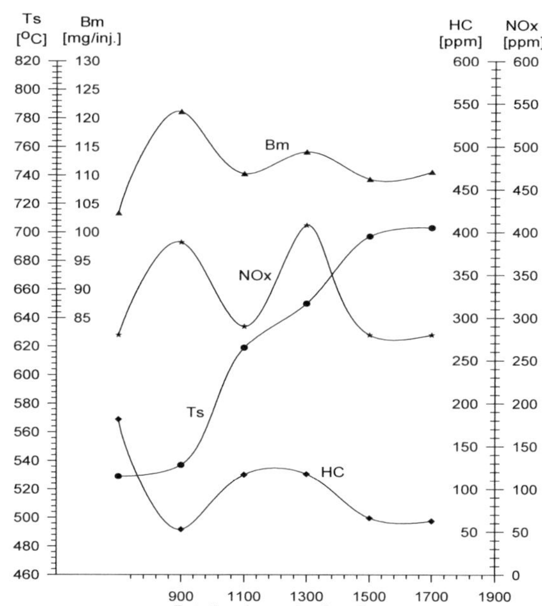

Figure 10. MD-111E engine parameters for WOT operation: Ts – exhaust temperature, Bm – fuel amount HC – hydrocarbons, NOx – nitric oxides

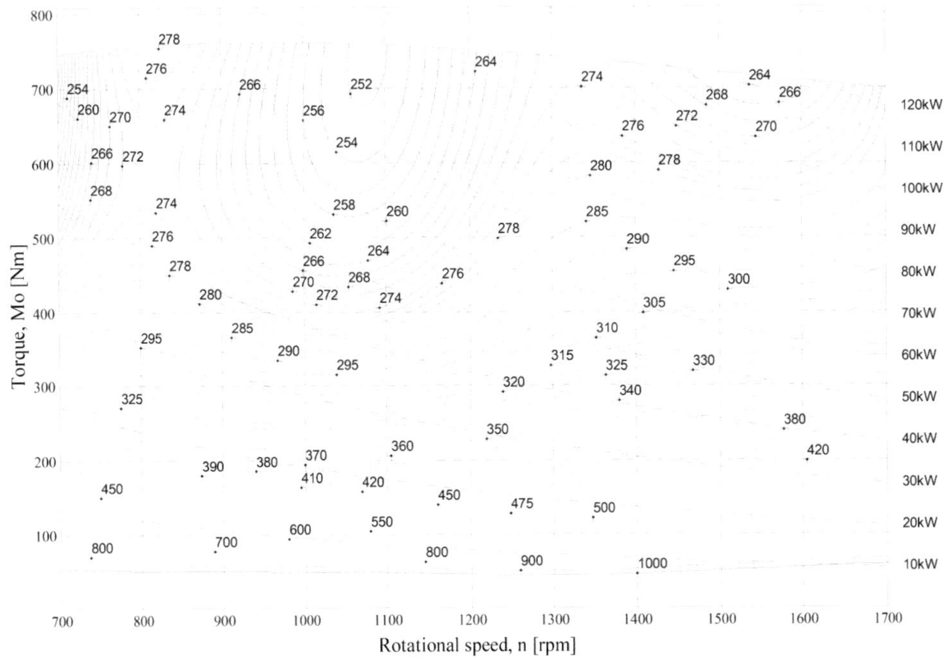

Figure 11. Generalized performance map of MDE-111E LPG engine equipped with sequential injection system and catalytic converter

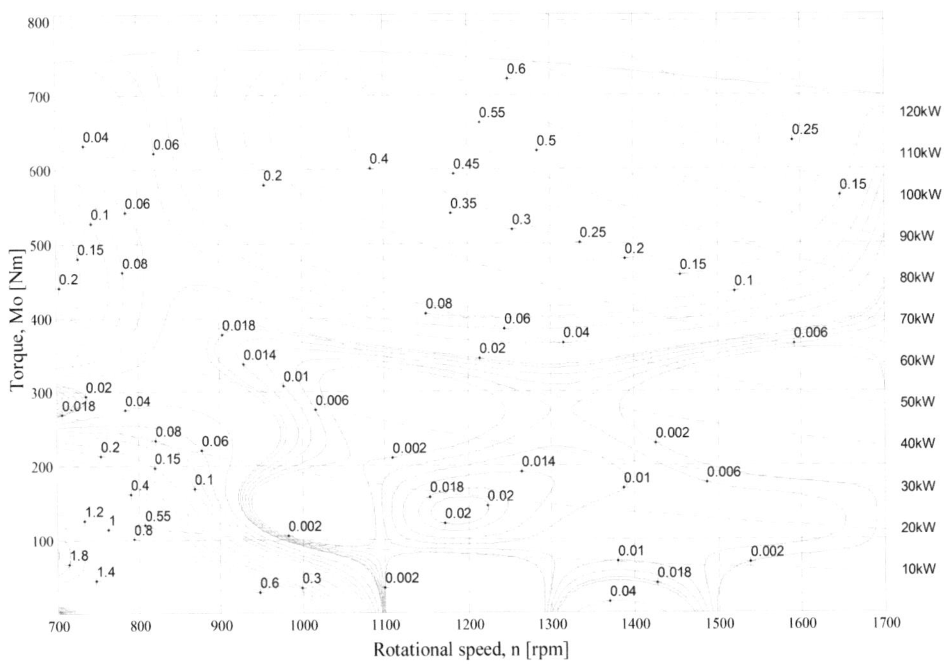

Figure 12. Contour map for concentration of monoxide carbon for MDE-111E LPG engine equipped with sequential injection system and catalytic converter

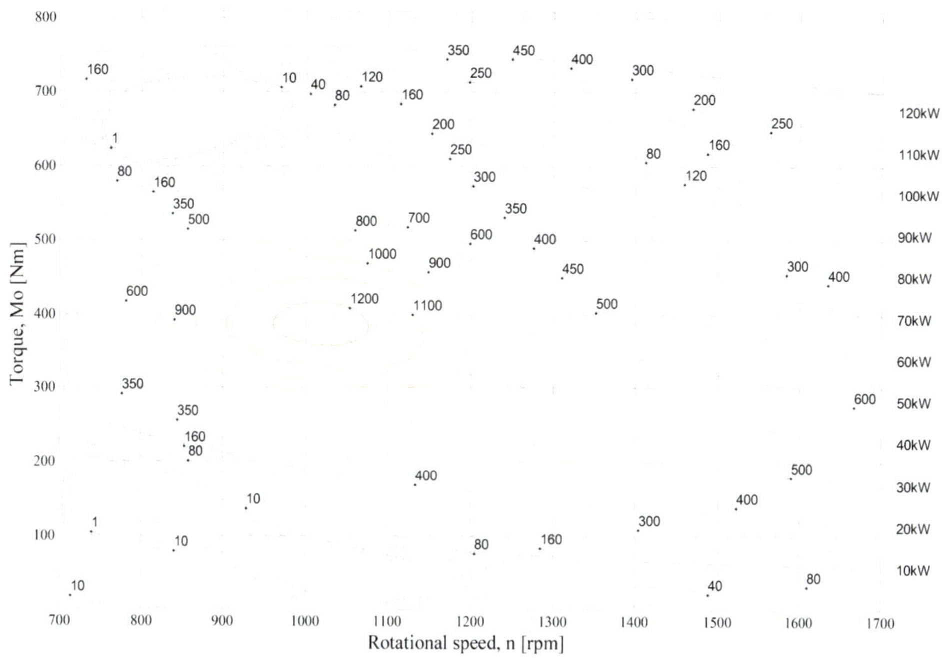

Figure 13. Contour map for concentration of oxides of nitrogen for MDE-111E LPG engine equipped with sequential injection system and catalytic converter

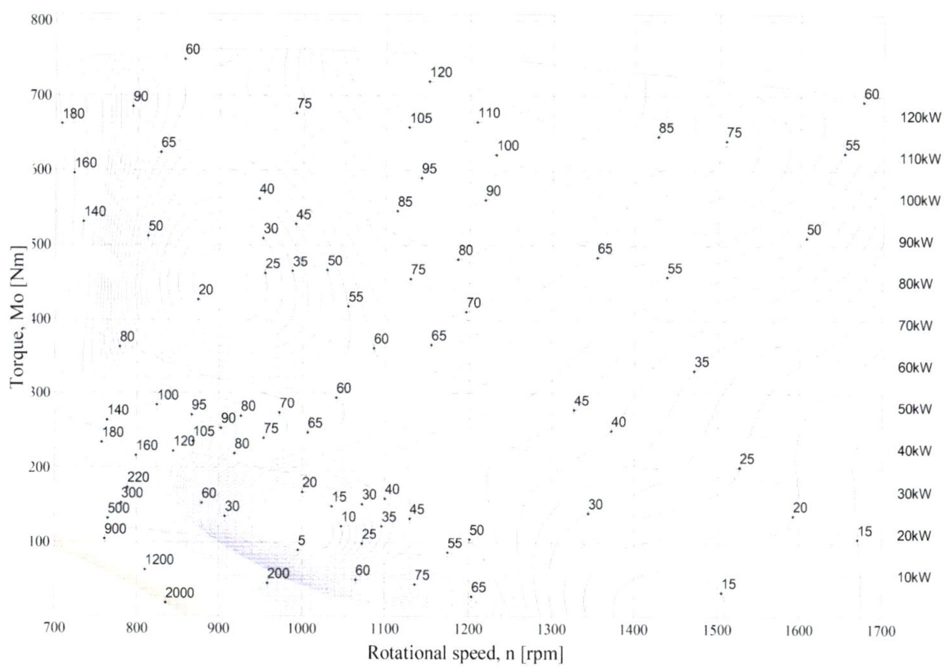

Figure 14. Contour map for concentration of hydrocarbons for MDE-111E LPG engine equipped with sequential injection system and catalytic converter

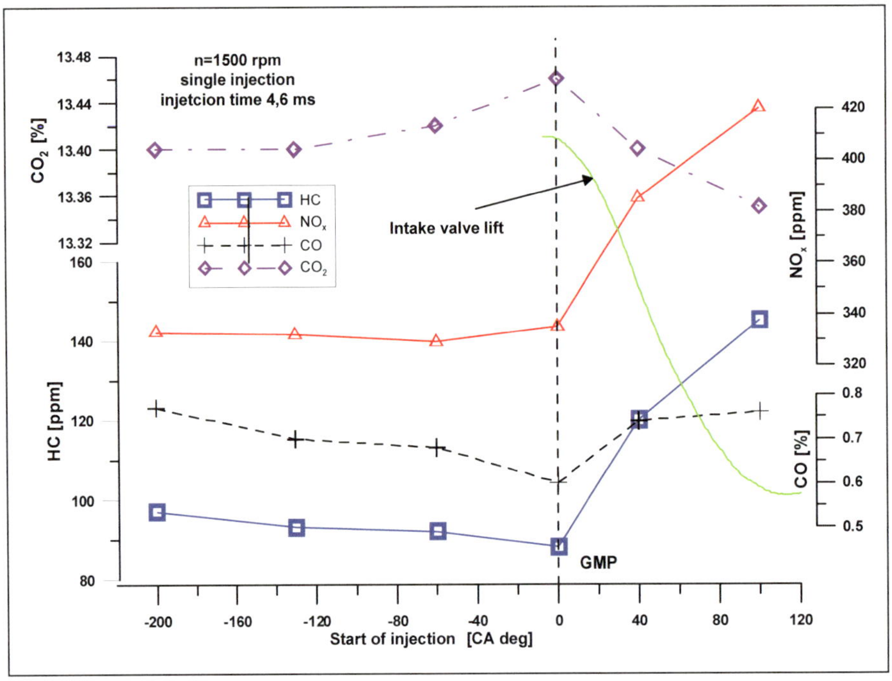

Figure 15. The effect of injection start on the concentration of CO, CO_2, HC, NO_x in the exhaust gas (single injection, n=1500 rpm, injection duration 4.6 ms) – for LPG in liquid phase

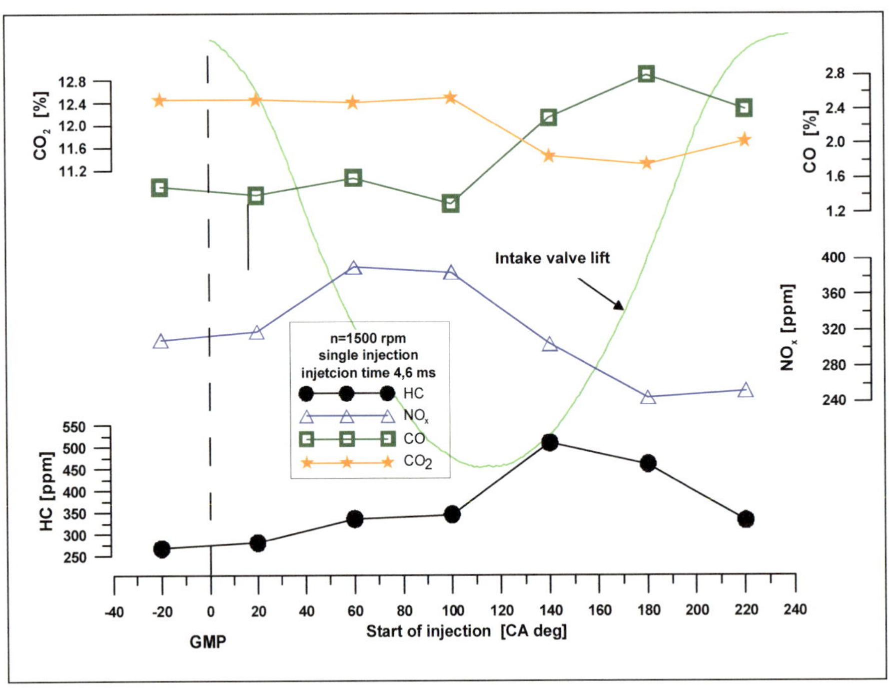

Figure 16. The effect of injection start on the concentration of CO, CO_2, HC and NO_x in the exhaust gas (single injection, n=900 rpm, injection time 5,5 ms) – for LPG in liquid phase

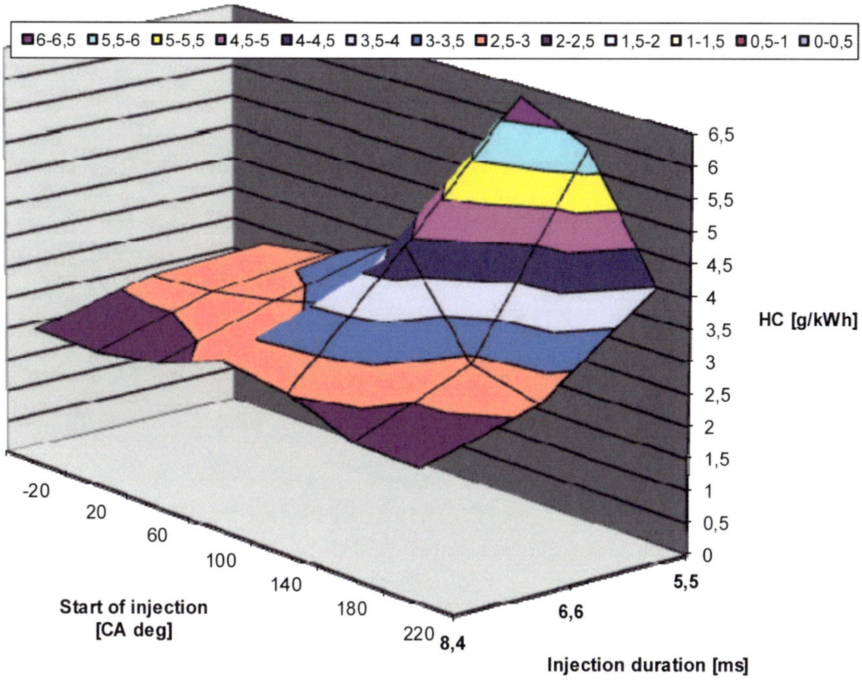

Figure 17. Specific HC emission for the selected injection parameters (single injection, n=900 rpm)

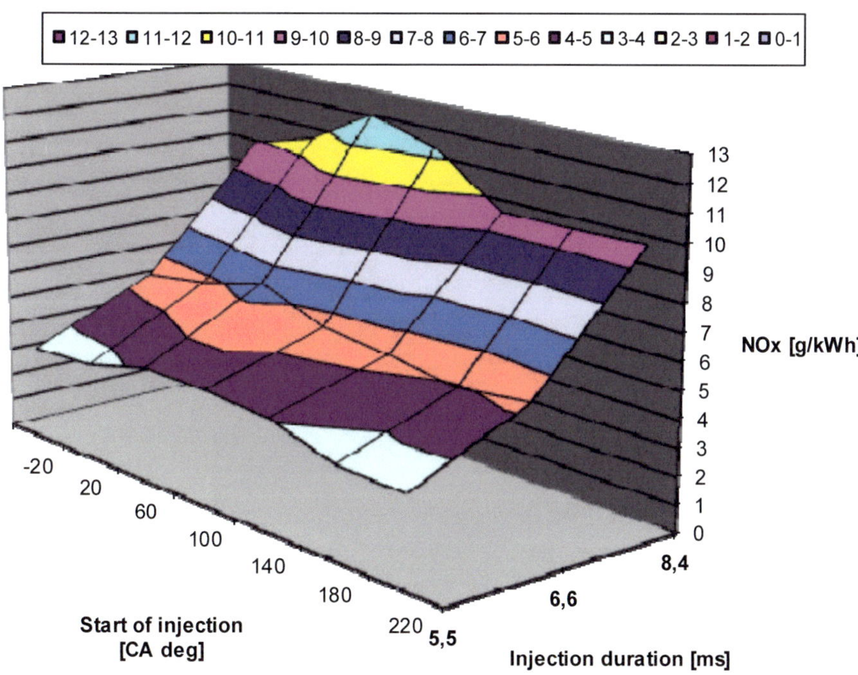

Figure 18. Specific NO_x emission for the selected injection parameters (single injection, n=900 rpm)

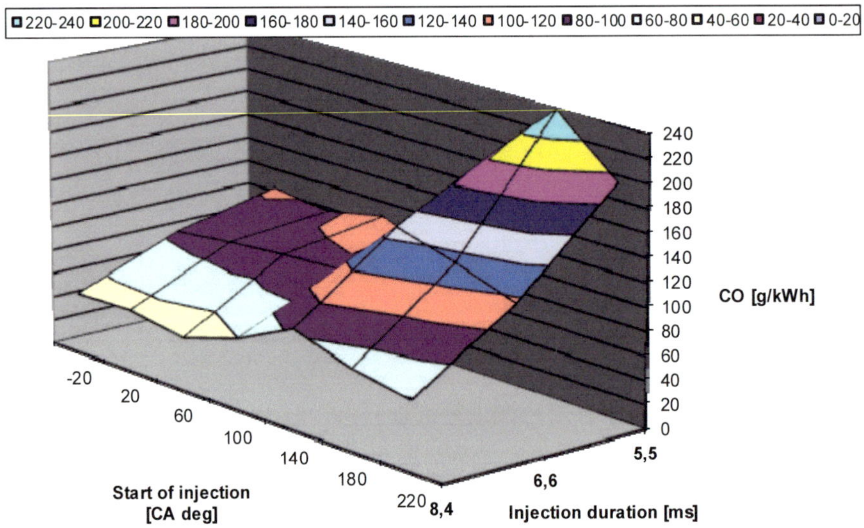

Figure 19. Specific CO emission for the selected injection parameters (single injection, n=900 rpm)

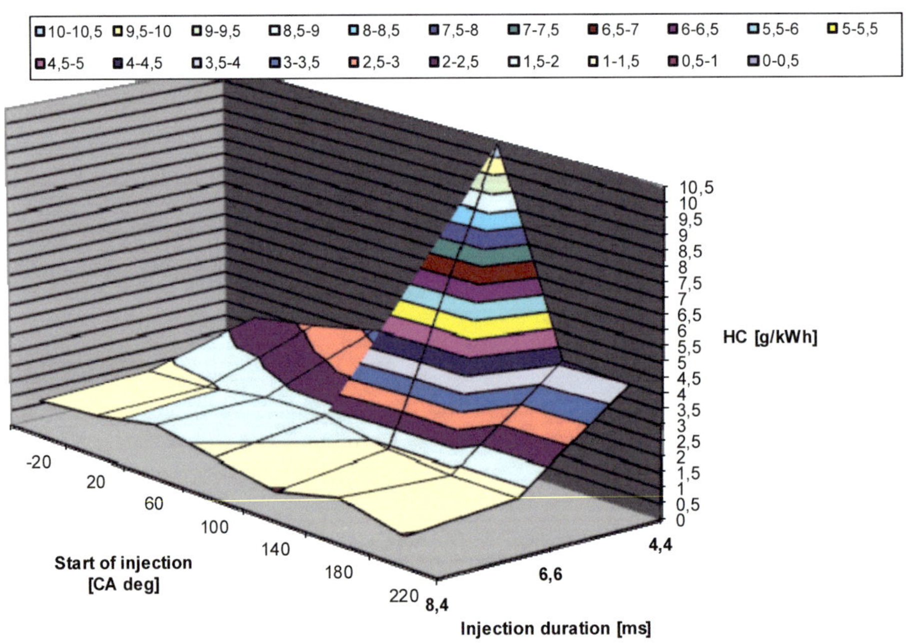

Figure 20. Specific HC emission for the selected injection parameters (single injection, n=1500 rpm)

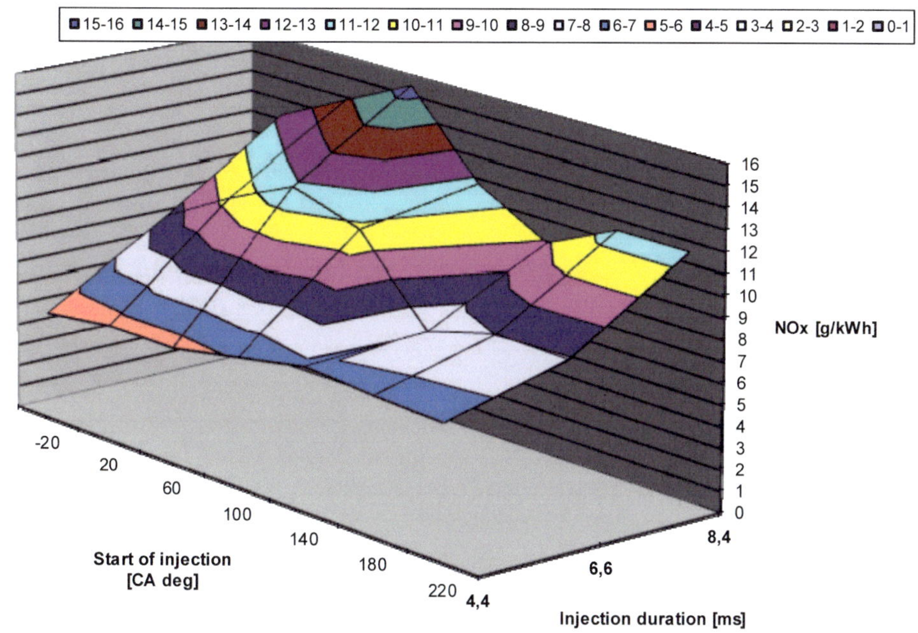

Figure 21. Specific NO_x emission for the selected injection parameters (single injection, n=1500 rpm)

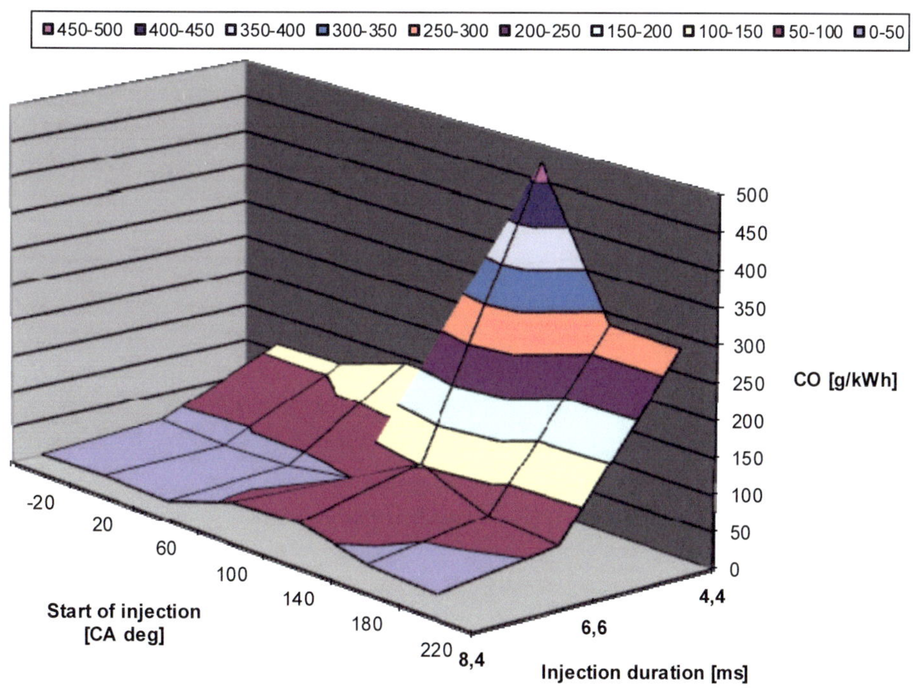

Figure 22. Specific CO emission for the selected injection parameters (single injection, n=1500 rpm)

At injection starts realized at closing the inlet valve the increase in specific CO emission is observed (fig. 19 and 22). The investigations show that specific CO emission decreases with injection duration (low load).

Table 2 presents the results of investigations on the effect of pilot and main injection on noise level generated by the engine. The study was conducted for an engine speed of n = 1500 rpm and three different loads – maximum, close to half of the maximum and not more than 10% of the maximum load. There was changed pilot injection advance α_{pp} relative to TDC (in the intake stroke), and the distance between pilot and main injection $\Delta\alpha_{pz}$. The study was conducted at a fixed value of ignition advance α_{wz} = 20 CA deg.

n [rpm]	M_o [Nm]	t_{inj} [ms]	α_{pp} [CA deg]	$\Delta\alpha_{pz}$ [CA deg]	α_{wz} [CA deg]	L_A [dB]	L [dB]
1499	10,7	4,4	285	100	20	98	102
1491	6,6	4,6	285	30	20	98	102
1498	7,7	4,4	285	160	20	98	102
1500	45,4	4,6	225	100	20	97	101
1501	336,0	10,2	285	100	20	98	102
1501	337,5	10,4	285	30	20	98	102
1500	329,2	10,4	285	160	20	97	101
1501	349,3	10,8	225	100	20	97	100
1500	735,0	18,0	285	100	20	97	100
1501	714,9	18,2	285	30	20	97	100
1501	713,8	18,2	285	160	20	97	101
1500	728,2	17,2	225	100	20	97	101

Table 2. The effect of pilot and main injection timing on nosie level of the engine

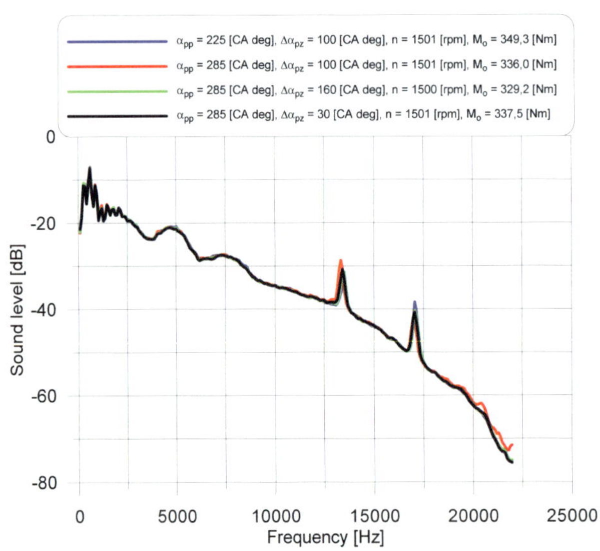

Figure 23. Frequency spectrum of sound level at various injection timing

From table 2 and fig. 23 we can see that at an engine speed of 1500 rpm, regardless of the load, both the start injection of pilot fuel quantity α_{pp} and the start injection of main fuel quantity (characterized by the value of $\Delta\alpha_{pz}$) do not have significant effect on the sound level and its frequency.

5. Conclusions

The tested engine MD-111E reaches more than 125 kW at maximum tested speed, but at a speed of 1500 rpm, when cooperates with a power generator has a power output of 115 kW, what allows to cooperate with a power generator with a capacity of 125 kVA providing a sufficient surplus of power.

The researches show, that injection timing has a significant relationship with the emission of toxic ingredients in exhaust gases of the engine. LPG fuel injection carried out at closing of intake valve causes an increase in specific HC emission and specific CO emission. For the injection starts realized at opening of the intake valve an increase in the specific NOx emission is observed. Realized researches show, that at an engine speed of 1500 rpm, regardless of the load, both the start injection of pilot fuel quantity and the start injection of main fuel quantity do not have significant effect on the sound level and its frequency.

The final value of timing and mutual location of the fuel quantities (for pilot and main injection) with respect to TDC of the piston can be selected only because of the optimal operation and environmental performances of the engine. It greatly simplifies the problem of optimization of the ignition system and fuel injection system designed to fuelling using LPG in liquid phase at sequential injection system and at split of injection.

The application of fuel system with dual LPG sequential injection of liquid phase and the catalytic converter can achieve satisfactory environmental performances of the engine. The use of a turbocharger gives the possibility to increase the engine power obtained in broad range of engine speed with small modifications of fuel system.

The use of broadband oxygen sensor instead of the two state oxygen sensor can improve the control accuracy and precision of fuel delivery. In this way can be met more and more rising requirements connected with emission standards.

Author details

Artur Jaworski, Hubert Kuszewski, Kazimierz Lejda and Adam Ustrzycki
Rzeszów University of Technology, Faculty of Mechanical Engineering and Aeronautics, Department of Automotive Vehicles and Internal Combustion Engines, Poland

6. References

Cipollone, R., & Villante, C. (2000). *A/F and Liquid-Phase Control in LPG Injected Spark Ignition ICE*. SAE Technical Paper 2000-01-2974

Cipollone, R., & Villante, C. (2001). A dynamical analysis of LPG vaporization in liquid-phase injection systems. *International Workshop on "Modeling, Emissions and Control in Automotive Engines" MECA'01*, University of Salerno, Italy, September 2001

Dutczak, J., Golec, K., & Papuga, T. (2003). Niektóre problemy związane z wtryskowym zasilaniem silników ciekłym propanem-butanem. *VI Międzynarodowa Konferencja Naukowa SILNIKI GAZOWE 2003*, Zeszyty Naukowe Politechniki Częstochowskiej, pp. 296-303, ISBN 83-7193-208-1, Częstochowa, 2003

Hyun, G., Oguma, M., & Goto, H. (2002). 3-D CFD analysis of the mixture formation process in an LPG DI SI engine for heavy duty vehicles. *Twelfth International Multidimensional Engine Modeling User's Group Meeting Agenda*, Detroit, 2002

Lee, E., Park, J., Huh, K.Y., Choi, J., & Bae, C. (2003). *Simulation of fuel/air mixture formation for heavy duty liquid phase LPG injection (LPLI) engines*. SAE Technical Paper 2003-01-0636

Oh, S., Kim, S., Bae, C., Kim, C., & Kang, K. (2002). *Flame propagation characteristics in a heavy duty LPG engine with liquid phase port injection*. SAE Technical Paper 2002-01-1736

PN-EN ISO 8178, part 1-4, 1999-2001

Training materials of Vialle. Kielce, 2001

Section 2

Engine Design, Control and Testing

Chapter 6

Intelligent Usage of Internal Combustion Engines in Hybrid Electric Vehicles

Teresa Donateo

Additional information is available at the end of the chapter

http://dx.doi.org/10.5772/48120

1. Introduction

Since early 1900s, gasoline and diesel internal combustion engines have represented the most successful automotive powering systems despite their low efficiency, their emissions issues and the increasing cost of fuel. Their main advantage over both gas engines and Battery Electric Vehicles (BEVs) is the very high energy density of liquid fuel that allows long driving ranges with small (and light-weight) storage tanks and safe and fast refueling processes. Moreover, gasoline and diesel fuels have an established infrastructure of distribution that is difficult and very expensive to replicate for other energy sources.

Environmental issues, energy crises, concerns regarding peaking oil consumption and the expected increase of number of cars in developing countries have eventually encouraged research into alternative energy sources. However, they are still unable to penetrate the market for several technological limitations.

The main drawback of BEVs resides in the batteries. They are still too expensive, too bulky and heavy (due to their low energy density). Moreover, they have an unsatisfactory life cycle and require long recharging times. Vehicles using fuel cell (FCV) a very clean fuel conversion system, have technologic drawback even higher. They add to the problems of a BEV, the use of a very light gaseous fuel that has severe limitations in terms of producing process, storing system, safety and distribution infrastructure. Thus, they are not to be considered as a viable way for eco-mobility in the next future (German, 2003).

Hybrid electric vehicles are characterized by the presence of two different typologies of energy storage systems: usually a battery and a gasoline or diesel fuel tank. HEVs have no limitation of range with respect to conventional vehicle and use the existing distribution infrastructure. The main advantages of HEVs are: the flexibility in the choice of engine operating point that allows the engine to be run in its high efficiency region and the possibility of downsizing the ICE and so obtaining a higher average efficiency. Moreover, the engine can be turned off when the vehicle is arrested (e.g., at traffic lights) or the power request is very low (reduction of the idle losses).

PHEVs can be considered either as BEVs that can be run in hybrid mode when the state of the charge (SOC) of the batteries is low or as HEVs with batteries that can be recharged from the electricity grid. They are characterized by the use of much larger battery pack when compared with standard HEVs. The size of the battery influences the All Electric Range (AER), an important design parameters of PHEVs that is defined as the number of miles they vehicle can run in pure electric mode on the UDDS cycle. A vehicle is classified as PHEVXY if it has an AER of XY miles.

PHEVs require fewer fill-ups at the gas station than conventional cars and have the advantage, over HEV, of home recharging.

BEVs, HEVs, and PHEVs have also the capability of partially recovering energy from brakes by inverting the energy flow from batteries to wheels through the electric machine.

Simpson, 2006 presented a comparison of the costs (vehicle purchase costs and energy costs) and benefits (reduced petroleum consumption) of PHEVs relative to HEVs and conventional vehicles. On the basis of his model, Simpson found that PHEVs can reduce per-vehicle petroleum consumption. In particular, reductions higher than 45% in the petroleum consumption can be achieved using designs of PHEV20 or higher (i.e. vehicles containing enough useable energy stored in their battery to run more than 20 mi (32 km) on the UDDS cycle in electric mode according to the previous definition of AER).

The study of Simpson, 2006 underlined that from the economic point of view, the PHEVs can become a competitive technology is the cost of petroleum will continue to increase and the cost of the batteries will decrease.

Because of different characteristics of multiple energy sources, the fuel economy and the environmental impact of hybrid vehicles mainly depend on a proper power management strategy. The particular operating strategy employed in this kind of vehicles significantly influences the component attributes and the value of the PHEV technology (Gonder et al. 2007).

Generally speaking, the environmental impact of an ecologic vehicle has to be determined with a "well to wheel" (WTW) approach. From a "tank to wheel" (TTW) point of view, a BEV, or a PHEV running in electric mode do not produce either pollutant or greenhouse gases while the emissions of pollutant and CO_2 in the WTW processes depend on the primary source and the technology used to generate electric energy at the grid. The well-to-wheel CO_2 emissions of a FCV can be equal to those of a diesel engine vehicle if it uses hydrogen produced from non-renewable energies sources (Guzzella and Sciaretta, 2007).

In a hybrid vehicle, the local emissions of CO_2 and pollutant strongly depend on the management strategy used for the ICE that becomes the main issue in both HEVs and PHEVs.

2. Classification of hybrid vehicles

Hybrid Electric Vehicles can be classified according to their architecture, the discharge/recharge mode of batteries and the level of hybridization.

As for architecture, HEV are called "parallel" when they use a gasoline or diesel engine mechanically coupled with an electric motor at the same shaft to satisfy the power request at the wheels. A parallel HEV can be run in five modes of operation (Guzzella et al, 2007): power assist (the electric motor give the supplementary torque to the shaft when the request is higher than engine available torque), battery recharging (a part of the engine power is used to recharge the batteries), electric mode (engine turned off), conventional vehicle (electric motor turned off) and regenerative braking.

In a "series" hybrid, the power request is entirely satisfied by the electric motor. Electric current to the motor is the algebraic sum of the current form/to the batteries and the current produced by an engine-driven generator. A series HEV can be run in four modes (the same of a parallel vehicle apart from conventional mode since the engine is not connected to the shaft).

Combined hybrid that can be run either in parallel and series mode have also been developed and introduced in the automotive market.

Traditionally, series HEVs have been neglected in scientific literature since they are less efficient than parallel HEVs and require more additional weight. Moreover, their energy management was considered trivial: a simple on-off engine control was considered sufficient. However, the increasing interest in plug-in vehicles has given new impulse to the research of advanced control strategies for series architectures.

There are two possible ways to regulate the energy management of hybrid vehicles with batteries. The first one (charge depleting mode, CD) accepts the batteries to be completely discharged during the mission. In this mode, the battery SOC can increase or decrease in time but it tends to be reduced along the mission. This approach can be considered for plug-in vehicles only. The second one (charge sustaining mode, CS) tries to keep the battery always charge to not affect the vehicle autonomy. The SOC can increase or decrease in time but it tends tore main constant during the mission (for series and parallel HEVs, not possible for BEV).

A PHEVs is usually run in CD mode without using the engine until reaching a pre-assigned lower bound on the SOC, then a CS strategy is adopted. Another possibility is to discharge gradually the battery throughout the trip as in the so-called *blended mode* control (Tulpule et al., 2009).

This makes a PHEV more complex, more dependent on traffic and route information and more efficient than a standard series HEV.

Another classification of importance for hybrids is the degree of hybridization. *Micro-Hybrids* are quite similar to conventional vehicles, from that they differ for the presence of a slightly larger battery and a little more powerful electric motor that allow the engine to be turned off when the car is stopped at the cross-lights and then turned on again when the vehicles moves. This system, named Start&Stop is nowadays adopted by several automotive companies in order to fulfill the Euro V standard. It does not require an increase of the voltage of the electric systems. The increase in cost and complexity is quite small like the potentiality to decrease fuel consumption. By further increasing the electric power and the voltage, it is possible to recover the braking energy (+5-10% in fuel economy, Chan, 2007). If the power of the electric motor increases, the internal combustion engine can be downsized and the electric motor is used to increase the pick power with the Power Assist logic. This is the case of *mild hybrids* (like Honda Civic and Honda Insight), that can increase fuel economy by 20-30% (Chan, 2007) with a similar increase of cost. Mild hybrids usually are not able to be run in all electric drive like *Full Hybrids*. Full hybrids can achieve a 40-50% higher fuel economy than conventional car. (Chan, 2007) They work with very high tension in order to accept the largest electric power. They can be sub classified in Synergy Hybrids and Power Hybrids. The former are designed to maximize fuel economy (downsized engine) while the latter use the electric motor to increase the available torque (no-downsizing).

Finally, the term *Range extender* is used to define series hybrid vehicles where the small engine-alternator group is only used to recharge the battery when their SOC is too low.

2.1. Designing and managing internal combustion engines for hybrid applications

The role of the internal combustion engine in hybrid electric vehicles (HEVs) is quite different from conventional vehicle. The engine has no more to be designed to fulfill the performance (maximum speed, acceleration and climb) required for the vehicle but can be downsized, thus reducing fuel consumption and greenhouse emissions. Moreover, the internal combustion engine can be better managed in order to avoid low-efficiency and high-emission operations like idling, vehicle stops and strong accelerations.

The current approach to HEV design is to use internal combustion engines developed for conventional vehicles. From this point of view, the advantage of fuel economy of HEVs can be actually defeated by the higher complexity, weight and volume of the power-train. However, many of the electronic-controlled devices used in engine to increase their efficiency and reduce emissions at idle and low speed-low torque operating mode are completely useless in HEV applications. This means that simpler, lighter and less costly engine could be developed for hybrid applications.

It is well known that internal combustion engines have poor fuel economy and larger if they work at low temperature. This is particularly important in hybrid electric vehicles since they allow the engine to be turned off for long periods during which the engine temperature decreases. This can lead to higher cold-start emissions particularly due to the poor conversion efficiency of the after-treatment devices when the light off temperature is not reached. On the other hand, hybrid electric allow either engine or after-treatment devices or both devices to be controlled to reduce the warm-up period and improve their performances in a fully integrated approach (Bayar et al., 2010). In HEV, the engine is cranked to higher speed than conventional vehicles and this makes the combustion condition during startup process quite different. Yu et al., 2006, investigated the effect of cranking speed on the start/stop operation of a gasoline engine for hybrid applications. Once again, fuel economy and emission during the engine start process depend on the control strategy used for the engine and the motor.

In order to reduce the warm-up period of the engine Lee et al., 2011 considered the recovering of exhaust gas heat exchanging system with coolant and gear box oil simultaneously. Accordingly, they developed an exhaust heat recovery device, which performs integral heat exchange of the exhaust gas heat of engine to increase the temperature of the coolant and the gear box oil, thereby reducing friction loss and improving fuel economy.

2.2. Approaches to the supervisory control models

The capability of a HEV in reducing fuel consumption and pollutant emissions strongly depends on the supervisory control strategy and the specific driving conditions. In fact, in hybrid electric vehicles a supervisor control system defines in each time the power split between the fuel conversion system (engine/alternator or fuel cell) and the electric storage systems (batteries and/or super capacitors) in order to minimize fuel consumption, sustain battery charge and reduce polluting emissions. Note that these goals are competitive and the performance of the HEV strongly depends on which goal it is given a higher importance. The optimization should be performed, ideally, over the entire life cycle of the vehicle even if a much shorter time interval (from a small number of minutes to few hours) is usually taken into account.

Several approaches for the optimization of energy management of a HEV have been presented in literature (Serrao, 2009). They can be classified in four categories: numerical optimization, analytical optimal control theory, instantaneous optimization and heuristic control techniques.

Heuristic control techniques are based on a set of rules that generate control action (i.e., the power to be delivered from the two energy sources) according to the value of some vehicle parameters like speed, acceleration, battery SOC, etc. These methods easy to implement in vehicles but they do not guarantee the minimization of either fuel consumption or emissions and the achievement of charge sustaining at the end of the mission.

Numerical optimization usually applies dynamic programming to optimize the vehicle behavior with the unrealistic assumption of perfect knowledge of the vehicle driving conditions (Lin et al, 2003).

An alternative to dynamic program is the application of the Pontrayagin's principle. This approach assumes that the power train can be described with simple analytical functions. Thus, it is often a too simplified approach and it also requires the knowledge of the driving cycle to be applied (Anatone et al. 2005, Serrao et al. 2008).

In the instantaneous optimization approach, the global minimization problem is implemented and solved as a sequence of local optimization problems. The best known of these strategies is the Equivalent Consumption Minimization Strategy for charged-sustaining vehicles. The ECMS tries to minimize the equivalent fuel consumption that is calculated as the sum, in a time interval Δt, of the actual engine fuel consumption and the fuel equivalent of the electric energy stored in/extracted from the battery in the time interval Δt. Since battery is only used as an energy buffer, its energy is produced ultimately by the fuel that the engine has consumed/saved in the past (or will consume in the future). The main drawback of the approach is that it requires the definition of equivalent factors in the conversion of fuel energy to electrical energy and vice versa (Guzzella and Sciarretta, 2007).

Recently, Millo et al. 2011 extended the ECMS technique to include engine emissions. In particular, they correlated the use of the battery with equivalent NOx emissions and compared the results of the fuel consumption-oriented optimization and the NOx optimization in terms of State of Charge history, engine operating points, etc. with respect to several standard driving cycles.

The usage of standard driving cycles in the optimization of the control strategies is a common way to obtain sub-optimal controller that, however, can give poor results in the real driving conditions.

2.3. Prediction of vehicle driving patterns

As explained before, the possibility of estimating the future driving profile (speed and related power demand) is a key issue in the development of hybrid vehicles. In fact, the supervisory controller of a HEV could use the future speed profile to optimize the power split in a future time window in order to minimize fuel consumption, pollutant emission, battery usage and so on. Moreover, the information about future can be used to activate the electric warming of engine and after-treatment devices. In this way they will be at the right temperature when the engine will be turned on and the exhaust gas flow will enter the after-treatment device.

In literature, a number of "auto-adaptive" techniques which try to predict future driving conditions based on the past ones have been defined A possible approach is to predict the future driving conditions based on the past behavior of the vehicle (Sciarretta et al, 2004) relying on the assumption that similar operating conditions will exist. But the future driving profile also depends on the instantaneous decisions which the driver will take to respond to the physical environment (driving patterns). Moreover, recent studies have shown that driver style, road type and traffic congestion levels impact significantly on fuel consumption and emissions (Ericson, 2000, Ericson, 2001). For these reasons, the control strategies proposed in some schemes (Won et al, 2005) incorporate the knowledge of the driving environment.

In the case of series HEV, the knowledge of the driving conditions have been found in literature to be less important than in the case of parallel hybrids (Barsali et al,. 2004).

In the case of plug-in hybrid electric vehicles, the control is more complex, strongly depending of the initial value of SOC and on the mission length, particularly if Blended Mode control methods are used. In fact, if the total trip was known, the best results would be obtained if the SOC would reach the lower value at the end of the trip (Karbowski et al. 2006). Gong et al. 2007, developed an Intelligent Transportation System that uses GPS information and historical traffic data do define the driving patterns to be used in the optimization. Donateo et al. 2011, have estimate numerically that the knowledge of the driving cycle in a future time window of 60s can improve fuel consumption in a series PHEV with Blended Mode control by 20%.

3. ICT and sustainable mobility

3.1. Intelligent vehicle technologies

According to Gusikhin et al. 2008, a vehicle can be defined as intelligent if it is able to sense its own status and that of the environment, to communicate with the environment and to plan and execute appropriate maneuvers. The first application of intelligent vehicle systems has been the increase of safety by providing driver assistance in critical moments. A combination of on-board cameras, radars, lidars, digital maps, communication from other vehicles or highway systems are used to perform lane departure warning, adaptive cruise control, parallel parking assistants, crash warning, automated crash avoidance, intelligent parking systems.

Markel et al. 2008 studied the effect of integration between an electrified vehicle fleet and the electric grid in order to increase the amount of renewable energy used to power the electric vehicles by optimizing the timing and the power of the charging processes during the day. Different communication protocols have been considered and compared by Markel et al. Intelligent Transport Systems like traffic management can have a direct effect on the emissions of CO_2 produced by the automotive floats (Dimitrakopoulos, 2011). According to Janota et al. 2010, Intelligent Transportation Systems can reduce consumption and emissions by acting on the vehicle (by monitoring and controlling the engine), on the infrastructure (reduction of number/duration of congestions and stoppage, optimization of intersection, cooperative systems to avoid congestions) and on the driver (planning of ecologic routes based on real-time information, support to driver for economic drive).

Recently, Information and Communication Technologies (ICT) techniques have been proposed for gathering information about the vehicle routes and road conditions that could allow the evaluation of the future power request of the vehicle over a large time window. ICT techniques can be used to estimate the future driving profile, suggest low consumption behaviors to the driver, propose alternative route, communicate the position and the status of electric recharging stations, etc. (Sciarretta et al, 2004).

Schuricht et al. 2010 analyzed two active energy management measures. The first one, uses advanced traffic light, and communication systems to support the driver during intersection approaching. The second one explores the uses of information and sensor sources from the traffic telematics for the predictive online optimal control of hybrid vehicles.

3.2. The CAR approach

The role of Intelligent Transport Systems in the improvement of PHEV performance and spreading of vehicles electrification is a research issue at the Center for Automotive Research at the Ohio State University. Starting from the awareness that traffic, weather and road conditions will be available in the next future through vehicle-to-vehicle and vehicle-to-infrastructure communications, the researchers at CAR emphasize the possibility this information in order to adapt the tuning of the energy management controller in HEVs, predicting the future driving profile, signaling the availability of recharge stations, predicting the route and generating statistical information for modifying pre-stored maps.

In the paper of Tulpule et al. 2011, the authors concentrated on the impact of the available data on the energy management in order to identify the most important factors on the actual fuel consumption of a PHEV. The factors analyzed in the investigation, named "Impact Factors", derive from both weather information (temperature and humidity) and traffic information (status of traffic lights, presence of pedestrian, road events in intra-city highway and inter-city highway). Their importance on the performance of the ECMS strategy were evaluated on the basis of a large amount of data acquired on a Toyota Prius converted to plug-in mode. The plug-in Prius has been run for a total of 60,000 miles in the campus area of the Ohio State University and several parameters like GPS information, temperature, fuel consumption, battery SOC, etc. were collected along with time and date.

To study the effect of the driving patterns, Gong et al. 2011 used a statistic approach to analyze real world profiles and derive information about average speed, speed limits, segment length, etc. These data were used to build a series of reference driving cycles by using the Markov chain modeling. The results of the investigation showed that the driving patterns have a relevant effect on the performance of a plug-in HEV and that the statistic values of acceleration have the largest impact of the tuning of the ECMS strategy.

3.3. The CREA approach

The CREA idea of intelligent hybrid vehicle includes the possibility of sensing the traffic environment in which it moves to predict the future driving conditions (Ciccarese et al. 2010). In particular, the vehicle is assumed to receive information from GPS, on-board sensors and vehicular communications. The scheme of the intelligent HEV according to the CREA research center is shown in Figure 1.

This information can be used on-board to perform a simulation of the traffic in a pre-set time window in order to predict the power request pattern in the next future and execute on-line optimization of the energy management over the predicted power pattern. The main difference with the CAR approach is that the vehicle is assumed to be able to compute on-board a simulation of the traffic conditions by using a microscopic road traffic simulation to derive its own future power request profile and optimize fuel consumption, battery usage, emissions levels, etc. This approach requires a relevant on-board computational capability that we believe could be available in the next future for other applications like safety, entertainments and so on. Alternatively, the simulation of the traffic patterns and the calculation of the speed profiles of the vehicles in a particular urban zone could be performed by a central computational unit that could send the results to the vehicles circulating in that zone.

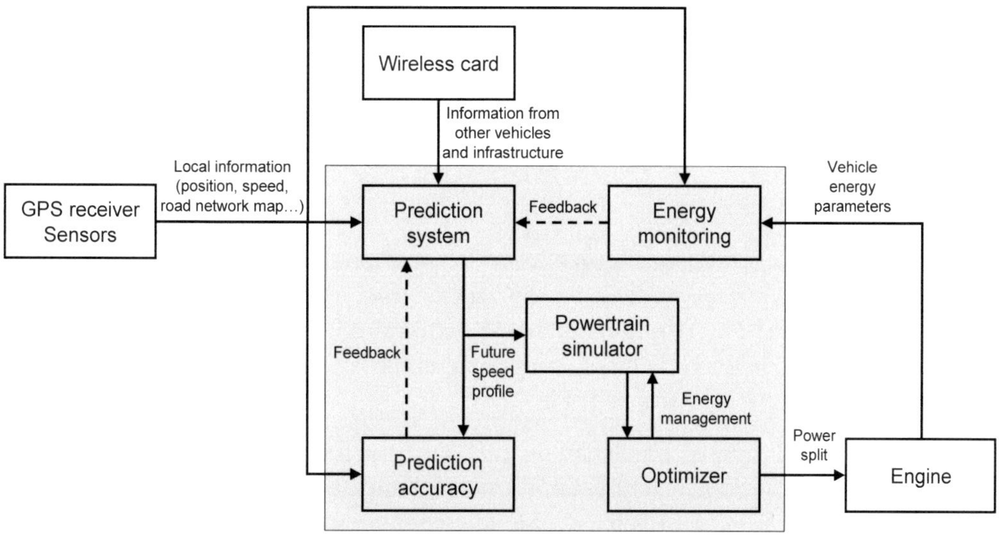

Figure 1. An intelligent hybrid vehicle according to the CREA approach

The gray area in Figure 1 represents the tools to be implemented on board. They include the prediction system, which is used to estimate the future speed profile of the vehicle, a power train simulator, which evaluates the evolution of fuel consumption and battery SOC during the prediction interval, and an optimizer, which is used to optimize the parameters of the control strategy.

3.3.1. The prediction block

This block gathers status messages that surrounding vehicles and/or the infrastructure broadcast. Messages transmitted by a vehicle carry status information, such as position, speed, acceleration, etc., and, optionally, some information related to its route. Messages generated by the infrastructure, instead, carry the current status and the timing of traffic lights. Besides the status information received through vehicular communications, the system gathers the status information on the "predicting vehicle" locally obtained by a GPS receiver and/or on-board sensors and also retrieves the data on road network from the digital maps used by the GPS navigation device.

The information gathered is exploited to take, at regular intervals, a snapshot of the traffic scenario in a given area. Each snapshot is the input to a run of module which simulates the traffic dynamics over a certain time interval, whose duration is at most equal to the prediction horizon. In Ciccarese et al. 2010, a modified version of SUMO software has been considered as on-board simulator.

SUMO (Simulation of Urban MObility) is an open source microscopic road traffic simulator. The input parameters of SUMO consist of the road network, the characteristics of each vehicle, the path (route) that each vehicle follows and the timing of traffic lights.

Vehicles with the same characteristics are grouped in classes and, for each class, a set of mechanical specifications is provided maximum speed, acceleration and deceleration., vehicle length, mass, friction coefficients, etc.

The road network is represented by an oriented graph, where nodes correspond to intersections and arcs to one-way lanes. For each lane, the maximum speed, the slope and the classes of vehicles which are allowed to go along it have to be also specified. The route of a vehicle consists of a list of consecutive arcs in the graph.

Using the input data, SUMO generates a mobility trace for all vehicles according to a Car-Following model (Wang et al. 2001): each vehicle tries to hold its speed close to the maximum one allowed for the current lane and decelerates if it is approaching either to an intersection or to another vehicle on the same lane; in the latter case, its speed is adapted to that of the vehicle which moves ahead of it.

The accuracy of the proposed prediction method has been tested experimentally (Ciccarese et al. 2012) in a augmented reality environment to simulate the presence in the Ecotecke campus of a certain number of vehicles able to communicate with the target vehicle. The experimental campaign showed that the inaccuracy of the prediction method is below 4km/h. In Figure 2, a comparison is shown between the predicted and the actual speed profile of the target vehicle in a time window of 100s. More details about the experimental campaign can be found in Ciccarese et al. 2012.

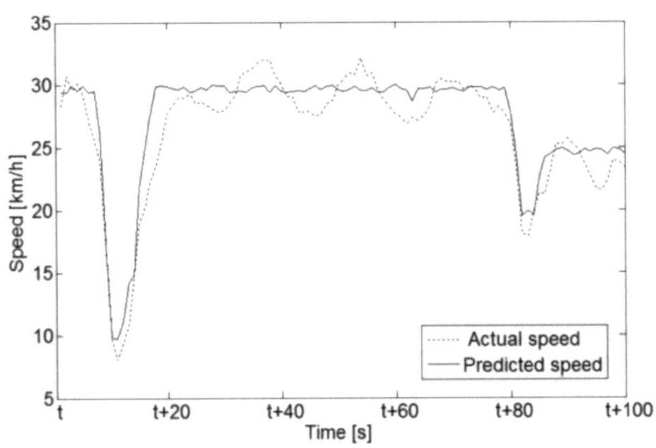

Figure 2. Example of speed profiles obtained by the experimental environment

3.3.2. *The power-train simulator*

The Power train simulator block implements a model of the power-train. The block processes the output of the prediction system and calculates the related power demand of the predicting vehicle by considering aerodynamic force, inertial contribution, rolling force and grade force. Information from on board sensors (ambient temperature, asphalt conditions, tires pressure and temperature) can be used to correct the predicted load. Then, the block simulates the energy flows according to the selected energy management strategy (described above) and evaluates the evolution of fuel consumption and battery SOC during the prediction interval.

Two different paradigms are usually considered to simulate a hybrid vehicle (Guzzella and Sciaretta, 2007). In the backward paradigm, the velocity of the vehicle is an input. According to the vehicle specification and speed values, the power request at the wheel is calculated. By means of static maps, the energy consumption of both engine and batteries is calculated according to the selected energy management strategy. If the power-train is not able to meet the cycle requirements, the acceleration is reduced and the vehicle diverges from the driving cycle.

In a forward or dynamic model, the power requested by the driver through the acceleration and braking pedals is used as input to evaluate the acceleration and the vehicle speed. This kind of model is used for the development of the control systems, while the backward method is best suited for analysis and evaluation of the energy and power flow in the vehicle driveline. Thus, a backward model is considered in the proposed scheme.

If the driving cycle is predicted with a traffic model that takes into account the actual acceleration and deceleration capability of the power-train, it is not necessary to check if the vehicle is able to follow the prescribed driving cycle.

3.3.3. The energy management system

This block implements the supervisor control system which defines, at each time, the power split between the fuel conversion system (engine/alternator in a series HEV) and the electric storage systems (generally batteries) with the constraints that the sum of the power extracted from each energy source must be equal to the total power requested at the wheels.

3.3.4. The optimizer

The role of the optimizer block is to adapt the parameters of the actual control strategy to the future driving conditions. This block can be implemented either as a on-line optimizer or as a memory device for loading optimized maps (Donateo et al. 2011).

3.3.5. Monitoring blocks

The system also includes a block, named Energy monitoring, which monitors the energy parameters of the vehicle (engine efficiency, level of gasoline in the tank, battery SOC, etc.) and evaluates the effectiveness in optimizing the energy management. This evaluation is carried out at regular intervals of duration equal to the prediction horizon.

Another block, named Prediction accuracy, evaluates the prediction error (based on a comparison between the actual speed profile evaluated by GPS and that estimated by the prediction system). The output of the Prediction accuracy block could be used to trigger a new prediction run.

4. A test case: ITAN500

In order to evaluate the effectiveness of the CREA approach in reducing fuel consumption of a plug-in HEV, a numerical investigation has been performed with respect to ITAN500.. ITAN500 a four-wheel vehicle prototype with a size comparable with that of a large scooter. ITAN500 can be classified as PHEV40 because its all-electric range is 40 miles on the UDDS cycle.

The vehicle was designed to reach a maximum speed of 90km/h in hybrid configuration with a mass of about 800 kg. By taking into account the overall transmission ratio (1/3.46) the DC motor was selected in order to generate a torque of about 27 Nm at the speed of 3560 rpm. A set of six lead acid batteries in series are used to produce the nominal voltage of 72V required to feed the electric motor. The choice of lead acid batteries was due to the need of reducing the vehicle cost. However, other kinds of batteries are currently under consideration.

A small gasoline engine with a maximum power of 9.9kW at 3600 rpm is used to extend the range of the vehicle. More details on the power-train (shown in Figure 3) can be found in a previous publication (Donateo et al. 2012).

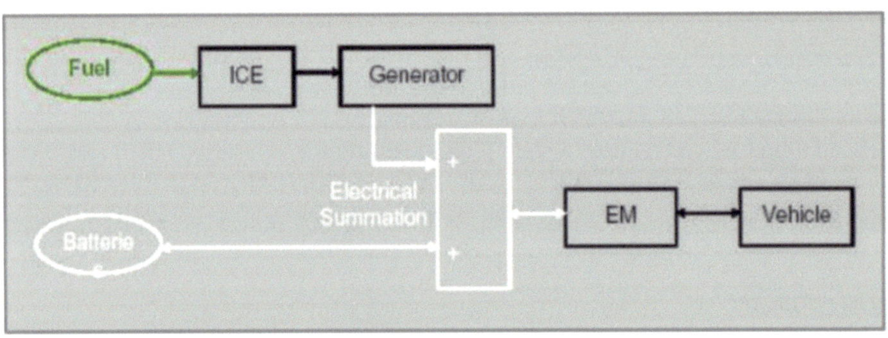

Figure 3. Scheme of the ITAN500 power-train

4.1. The VPR power-train simulator

VPR (Vehicle Power Request) is a backward model that uses quasi-static maps for the main power-train components (thermal engine, motor and batteries) to predict their efficiency according to the requested values of torque and speed.

The main outputs of the VPR model are the evolution of fuel consumption and battery SOC along the driving cycle. Starting from the velocity speed and grade traces, the vehicle power request is calculated by considering aerodynamic force, grade force, inertial contribution and rolling force. An example of vehicle power request trace is shown in Figure 4 together with other VPR output.

Note that during deceleration the power request is negative which means that the braking energy can be recovered and stored in the batteries. In the example shown in Figure 4, engine is turned on only in a small fraction of the vehicle missions, around 500s from the start of the cycle.

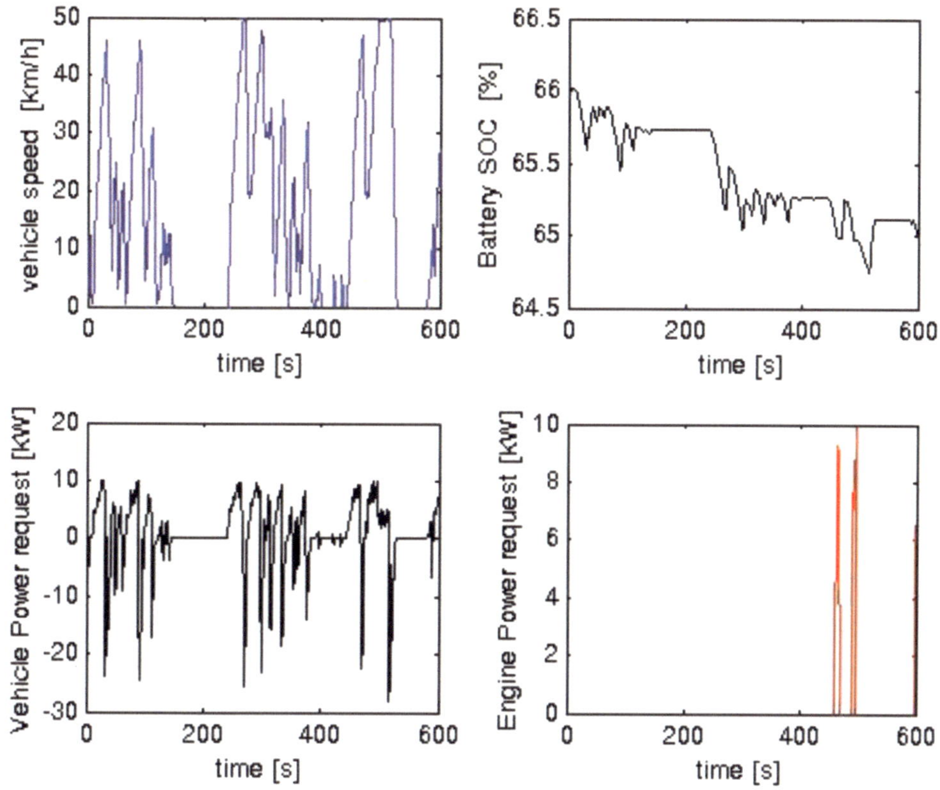

Figure 4. Example of VPR results

The efficiency of the electric motor according to torque and speed has been evaluated experimentally on an inertial test bench (Donateo et al. 2011).

Since the engine is run at the constant speed of 3000 rpm, its efficiency is considered as function of torque only. Literature data have been used to derive the maps of Figure 5.

The data of Figure 5 refers to a fully-warmed case, i.e. the temperature of the engine block is at the nominal temperature of 90°C. However, engine efficiency is strongly dependent on its temperature; in particular it is very low at cold start. In the VPR model, the efficiency data of Figure 5 are corrected as proposed by Guzzella et Onder, 2004 by multiplying the full-warmed engine efficiency by a correction factor whose dependence on temperature is shown in Figure 6.

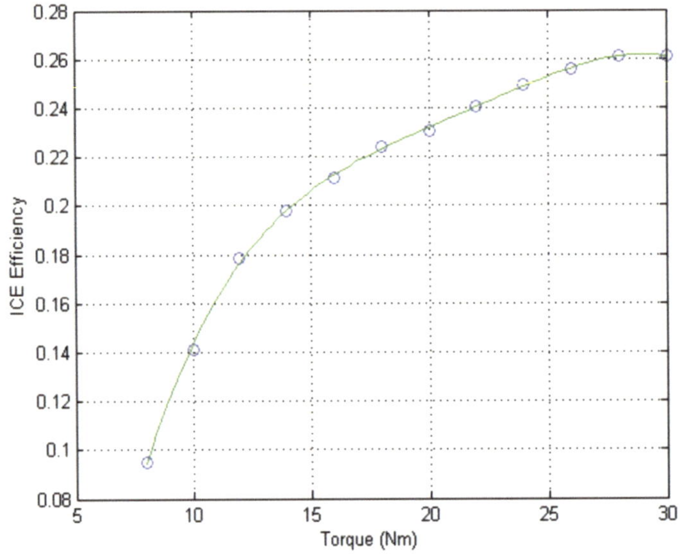

Figure 5. Fully-warmed engine efficiency versus torque at 3000rpm

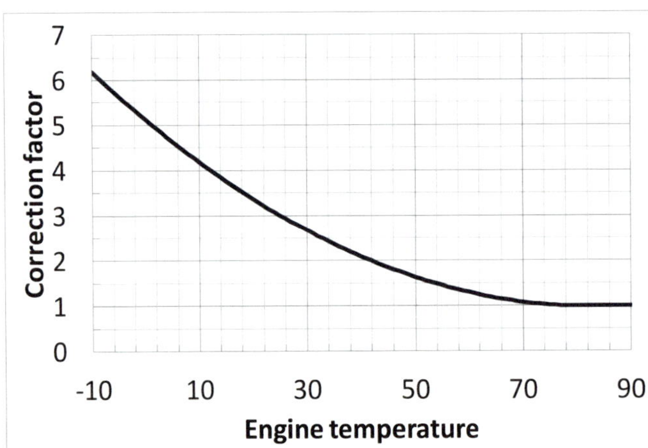

Figure 6. Correction factor for engine efficiency as a function of the block temperature

Note that if VPR is run on-board, the temperature of the engine block is a measured data while in the present investigation it has to be simulated. For this reason, a thermal model based on a zero-dimensional simulation of the engine has been proposed (Donateo et al. 2012). The thermal model is able to simulate the increase of temperature when the engine is on as a function of its actual torque. When the engine is off, its temperature decreases due to the heat transfer to the surrounding air. More details on the thermal model can be found in Donateo et al. 2012.

An example of temperature trace versus time obtained from VPR with the same input conditions of Figure 4 is shown in Figure 7.

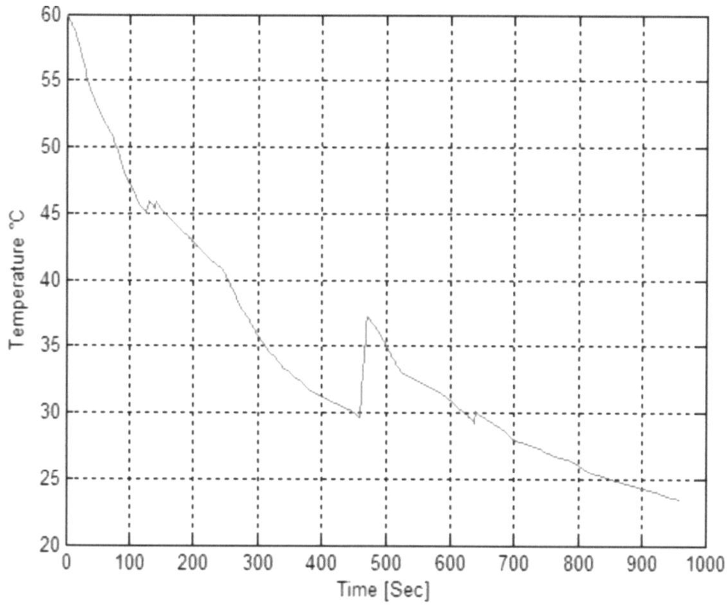

Figure 7. An example of temperature trace obtained with the thermal model of the engine

The overall electric efficiency between the chopper and the wheels is set constant and equal to 0.65 for the present investigation.

4.2. The energy management strategy

The energy management strategy developed for ITAN500 includes an initial Charge Depleting (CD) mode where the battery only is used until a threshold value of battery SOC is reached (SOC_{CD}). Then, the vehicle can be run in three different modes.

In Mode 1, the power to the motor is supplied only by the generator/engine group.

Mode 2 uses only battery to supply power. Both engine and battery are used in the other modes. In particular, in mode 3 the engine is used both to charge battery and to supply power to the motor while in mode 4 the engine and the battery are used together to feed the motor.

According to the actual power to be supplied to the motor to move the wheels (P_{load}) and instantaneous value of SOC, the power-train is operated in one of the Areas 1-11 of the Figure 8. In particular, mode 1 is preferred in the high power region except when the battery SOC is very high (Area 5). Mode 2 is mandatory in three cases: when the battery is fully recharged (Area 3), in braking (Area 1) and when the load power is very low (Area 2). Moreover, the use of mode 2 is preferred when the SOC is reasonably high and the load power relative low with respect to the engine nominal power (Areas 4 and 7), otherwise the use of engine is preferred (Areas 8). Area 6 and 9 correspond to the use of the engine to recharge the battery (mode 3). However, this is possible only when the sum of the load power and the power request to recharge the battery is lower than $P_{ICE,max}$. If not, mode 1 is used.

Note that Areas 11 and 10 of Figure 8 were not taken into account because the power request is always lower than $P_{ICE,max}$ for all the operating conditions considered in the present investigation.

The actual size of each area depends on the values of energy management parameters SOC_{CD}, SOC_{min}, k and $P_{ICE,min}$ that influences the results in terms of fuel consumption and battery usage over a specific vehicle mission. The meaning of SOC_{min}, and $P_{ICE,min}$ is quite straightforward while some explanation has to be given for K. The K parameter was introduced to solve the dilemma between using mode 1 or mode 2 in Areas 7 and 8 since neither the engine nor the battery works at their best in that region. By using K it is possible to prefer the battery at relative low power and high SOC (area 8) and the engine otherwise .

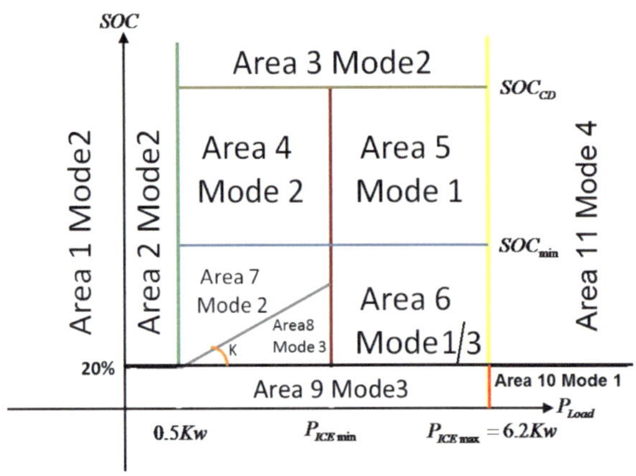

Figure 8. Energy management strategy

4.3. The optimizer

The optimal combination of the parameters can be easily performed off-line with a general optimization algorithm like genetic algorithms (Paladini et al. 2007).

The role of the optimizer is to find the optimal combination of parameters in Table 1 that define the size of the areas of Figure 8. For the optimization described in this paragraph, the minimum and maximum values and the steps of variation of the design variables reported in Table 1 were considered.

Variable	Min	Max
SOCCD (%)	60	80
SOCMIN (%)	20	60
K	0	1
PICE,min [W]	500	6200

Table 1. Design variables for the optimization

In each case, the goal of the optimization was the reduction of the equivalent fuel consumption calculated in the following way:

$$\dot{m}_{tot} = w_{FC}\dot{m}_{ICE}(\vartheta) + \dot{m}_{eq,BATT} \tag{1}$$

where:

$\dot{m}_{ICE}(\vartheta)$ is the effective fuel consumption function of engine temperature θ;

w_{FC} is the weight assigned to the level of fuel stored in the tank. It is set equal to 1 if the tank level is greater than 25% When the tank level is very low, this parameter is increased to prefer battery usage when the fuel level is low. In particular w_{FC} is 1.2 for 10%<tank_level<25% and 1.5 for tank level lower than 10%.

Note that eq. (1) has been obtained by adapting the equivalent fuel consumption defined by Sciarretta et al. 2004 for a parallel HEV to the specific power-train of ITAN500.

The equivalent fuel consumption of the battery is obtained as follows:

$$\dot{m}_{eq,BATT} = \frac{\eta^{\gamma} \cdot P_{BATT}}{Q_{LHV} \cdot \Delta t} \tag{2}$$

where η represents the average fuel consumption of the battery which is assumed to be constant and the same in charge and discharge in the present investigation.

When the battery is in charge, P_{BATT} represents the power that could be stored in the battery. Due to the battery efficiency η, the actual power stored in the battery (which define the equivalent fuel consumption) is lower than P_{BATT}. This is taken into account by setting γ =1. In discharge, P_{BATT} is the power requested from the battery is increased by η (γ=-1).

To complete the description of eq. (3), Q_{LHV} is the lower heating value of the fuel (in the present investigation gasoline is considered with Q_{LHV} =44MJ/kg while Δt is the time step of the driving cycle (Δt =1s).

The penalty function $f_p(SOC)$ takes into account the battery usage in the optimization process and has been defined according to Sciarretta et al. 2004.

4.3.1. *Driving cycles*

In the present investigation three kinds of driving cycles were taken into account for ITAN500. The first two are standard driving cycle adopted for the registration on new cars (NEDC and UDDS). Other numerical cycles were obtained with the help of SUMO. The ITAN500 has been simulated to move in the Ecotekne campus of the University of Salento for about 10000s (2.8h) together with other vehicles that, unlike ITAN500, can enter and exit the campus area. Different driving scenarios were taken into account by changing the number and the specification of the vehicles moving in the area.

The specification of the vehicles are used in the framework of SUMO to calculate the maximum values of acceleration/deceleration allowed to each vehicle according to the difference between the actual power request (depending on aerodynamics, rolling and inertia) and the maximum traction/braking power of the vehicle. Cycles obtained in this way were named as Trace A to Trace H. More details on the procedure used to obtain the numerical cycles can be found in Donateo et al. 2011.

Another cycle named R has been taken into account. This cycle is an actual driving cycle acquired with a GPS system on board of the vehicle ITAN500 when it is run in all electric range. The cycle has been assumed to be executed for 25 times (R*25) in order to obtain results of fuel consumption and battery usage comparable with those of cycles A and B.

The specifications of the cycles taken into account in the investigation are reported in Table 2. Note that all the cycles taken into account in the present investigation refer to a zero grade condition.

Cycle	Total time [s]	Average speed [m/s]	Max speed [m/s]	Min speed [m/s]
Cycle_NEDC	1225	8.93	33.36	0
Cycle_UDDS	1370	8.73	25.37	0
Cycle_1015	661	6.90	19.45	0
Cycle_HWFET	766	21.56	26.80	0
Trace A	10001	4.69	13.90	0
Trace B	10801	6.88	13.90	0
Trace C	9999	1.79	8.33	0
Trace D	10001	2.00	8.33	0
Trace E	10001	1.38	8.33	0
Trace F	10001	1.08	8.33	0
Trace G	10001	1.95	8.33	0
Trace H	10001	1.47	8.33	0
Cycle R	382	25.75	41.61	0.2

Table 2. Specification of the driving cycle taken into account for the creation of the maps

4.3.2. Full knowledge approach

In this approach the driving cycle is assumed to be completely known and the parameters of Table 1 are optimized for each cycle of Table 2. The results of the application of this approach to cycles A, B, R, NEDC and UDDS are reported in Table 3.

Cycle	**Duration [s]**	**Equiv.fuel cons. [l/100km]**	**Δ SOC [%]**	**FC [l]**	**SOC_{CD} (%)**	**SOC_{MIN} (%)**	**K**	**$P_{ICE,MIN}$ [kW]**
#A	10000	2.78	24.8	1.02	65	44.4	0.9	3.2
#B	10800	3.1	24.7	1.96	77.9	50.1	0.6	2.6
#R*25	9550	3.38	24.8	1.91	77.3	34.8	0.98	2.4
#UDDS	1370	1.66	18.7	0.08	60.6	37.6	0.97	6.1
#NEDC	1225	2.52	17	0.16	71.4	35.6	0.26	5.9

Table 3. Results of the optimization in the case of full knowledge (initial SOC 45%)

4.3.3. No-knowledge approach

The driving cycle is assumed to be completely unknown. The parameter of the control strategy are optimized for the NEDC cycle and applied to the other cycles. The results are reported in Table 4.

Cycle	Duration [s]	Equiv.fuel cons. [l/100km]	Δ SOC [%]	FC [l]	SOC_{CD} (%)	SOC_{MIN} (%)	K	$P_{ICE,MIN}$ [kW]
#A	10000	3.47	25	1.29	71.4	35.6	0.26	5.9
#B	10800	3.85	25	2.38	71.4	35.6	0.26	5.9
#R*25	9550	3.89	25	2.19	71.4	35.6	0.26	5.9
#UDDS	1370	1.67	18.7	0.08	71.4	35.6	0.26	5.9
#NEDC	1225	2.52	17	0.16	71.4	35.6	0.26	5.9

Table 4. Results of the optimization in the case of no knowledge (initial SOC 45%)

4.3.4. Prediction & maps supervisory control

In order to reduce the on-board computational load required by the CREA approach, Donateo et al. 2011 proposed the use of maps that are optimized off-line with respect to reference driving conditions. They were obtained with the following procedure.

All cycles of Table 2 have been taken into account to generate one global driving cycle of 85208s (about 23 hours). Then, the VPR has been used to calculate the corresponding power request according to the specification of the vehicle and a global power request trace has been obtained. This power request trace has been divided into 1420 Mini Power Cycles (MPC) of 60s.

The 1420 MPCs have been distributed in 90 groups with the help of the K-Means clustering technique. For each group, a representative driving cycle, named Reference Mini Power Cycle has been identified and numbered.

Figure 9 shows, with different colors, five MPCs belonging to the same group. The bold blue line is the RMPC chosen with the clustering algorithm.

The off-line optimization has been performed for each of the 90 RMPCs, two levels of engine temperature (cold-hot), three levels of the initial state of charge, and three levels of the fuel tank. In this way 1620 optimized maps have been obtained. Each map contains the optimized values of SOC_{min}, k and $P_{ICE,min}$. for a particular combination of RMPC, engine temperature, initial state of charge and level of the fuel tank.

The maps could be used in an intelligent hybrid electric vehicle in the following way.

1. At any interval of 60 seconds, the predicted speed profile is obtained from the prediction block;

2. The corresponding power request profile over 60s is calculated according to the vehicle and road specification (es. grade) with VPR;
3. The power request profile is compared with each of the RMPCs and the most similar one in terms of root mean square error is found;
4. According to the measured values of engine temperature, fuel tank level and battery state of the charge, the corresponding map is loaded;
5. The optimized values of the energy management parameters of the selected map are applied over the next 60s.

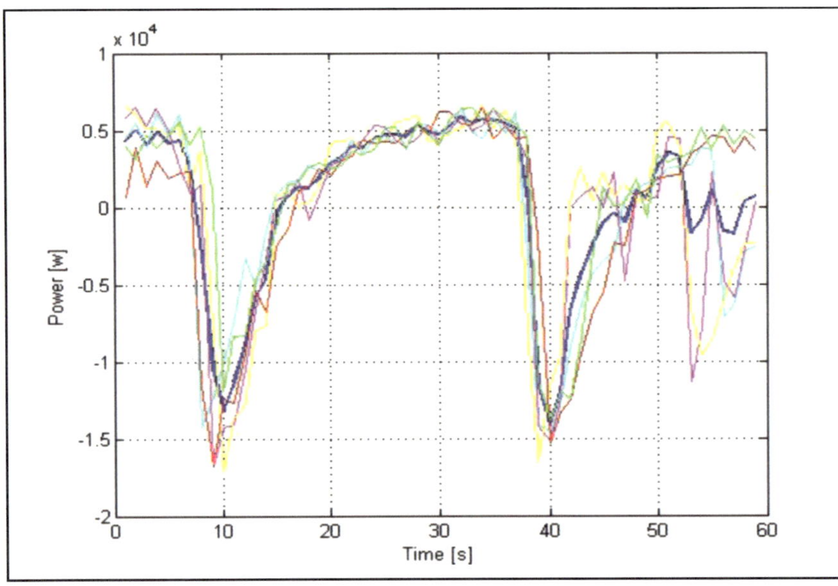

Figure 9. Example of RMPC

4.4. Analysis of the prediction&maps approach

The proposed on board prediction-optimization tool has been evaluated numerically in the following way. The ITAN500 is simulated to execute one of the driving cycles of Table 2 with the assumption that they are know (by prediction) in blocks of 60s.

At any 60s, the power request versus time in the next time window of 60s is evaluated with VPR and compared with each of the RMDCs to find the most similar one. Then, the instantaneous values of engine temperature, SOC and fuel levels are set as initial values and the corresponding optimized map is loaded. The thermal model of VPR is used to predict the profile of engine temperature along the mission. The values of the energy management parameters are used to evaluate the fuel consumption and battery usage in the next 60s on the basis of the actual power request (not on the selected RMDC).

The results in terms of fuel consumption and battery usage obtained with this approach are reported in Table 5.

Cycle	Duration [s]	Equiv.fuel cons. [l/100km]	ΔSOC [%]	FC [l]	SOC_{CD} (%)	SOC_{MIN} (%)	K	$P_{ICE,MIN}$ [kW]
#A	10000	3.38	24.9	1.26	From maps			
#B	10800	3.76	24.9	2.32				
#R*25	9550	3.73	25	2.09				
#UDDS	1370	2.42	13.4	0.19				
#NEDC	1225	2.61	15.6	0.18				

Table 5. Results of the simulation with the optimized maps (initial SOC 45%)

4.4.1. Percentage of mission with Controlled Battery Discharge (CBD%)

To compare the results of the three approaches, different metrics can be taken into account.

The first metric useful to compare the results of the three approaches can be derived by analyzing the typical SOC trace versus time in a plug-in hybrid electric vehicle. An example is shown in Figure 10 with respect to two different initial values of the battery SOC.

The traces of SOC show an initial zone where the results corresponding to *full knowledge, prediction&maps* and *no knowledge* are perfectly overlapped and the SOC decreases monotonically (Electric Mode). Of course this region is particularly evident and relevant when the initial SOC is higher (75%).

Then, there is a region in which the SOCs tends to decrease but can be kept locally constant or be increased thanks to the use of the engine (Plug-in Hybrid Mode). This region ends when the battery is fully discharged (SOC=20%). After this, the SOC remains globally constant for all cases (*full knowledge, prediction&maps* and *no knowledge*) with small variation that are not visible in the scale used for the Figures (Discharged Battery Mode). Thus, the different results in terms of fuel consumption obtained with the three methods can be accounted for with the different duration of the EM, PHM and DBM zones.

In the EM region, the fuel consumption is zero but the SOC strongly decreases due to the extensive use of the battery. In the PHM mode, the battery is the main energy source and the engine is turned on (when its efficiency is high) to decrease the slope of the SOC trace. The DBM region is the worst in terms of fuel consumption because engine has to be run also in its low efficiency region since batteries are fully discharged. A plug-in HEV is run at its best when the DBM region (SOC=20%) is reached exactly at the end of the mission and the EM region extends through as much of the mission possible. This is possible when the vehicle mission is entirely known (*full knowledge* case). The traces of Figure 10 show that the proposed method performs better than the *no knowledge* case since it allows to reduce the length of the DBM and to increase the PHM. As a consequence, the ICE is averagely run at high efficiency.

Thus, a useful metric to evaluate the performance of an energy management strategy for PHEV could be the percentage of the mission run in EM+DBM modes. This metric is named here CBD% while in a previous investigation (Donateo et al. 2012) it was referred to as Δmission.

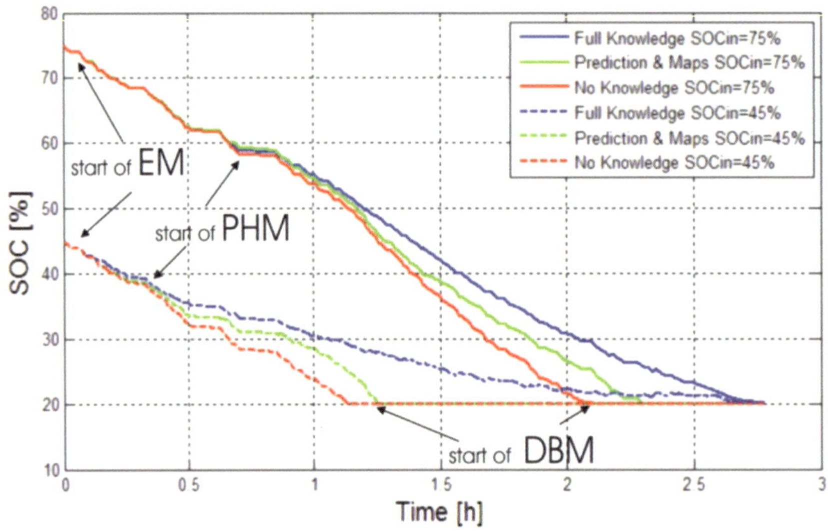

Figure 10. Explanation of the meaning of CBD% for Cycle A

The value of the CBD% has been calculated for each approach with reference to cycle A, B and R*25 of Table 2. R*25 means that cycle R has been repeated 25 times to achieve a duration similar to that of cycle A and B.

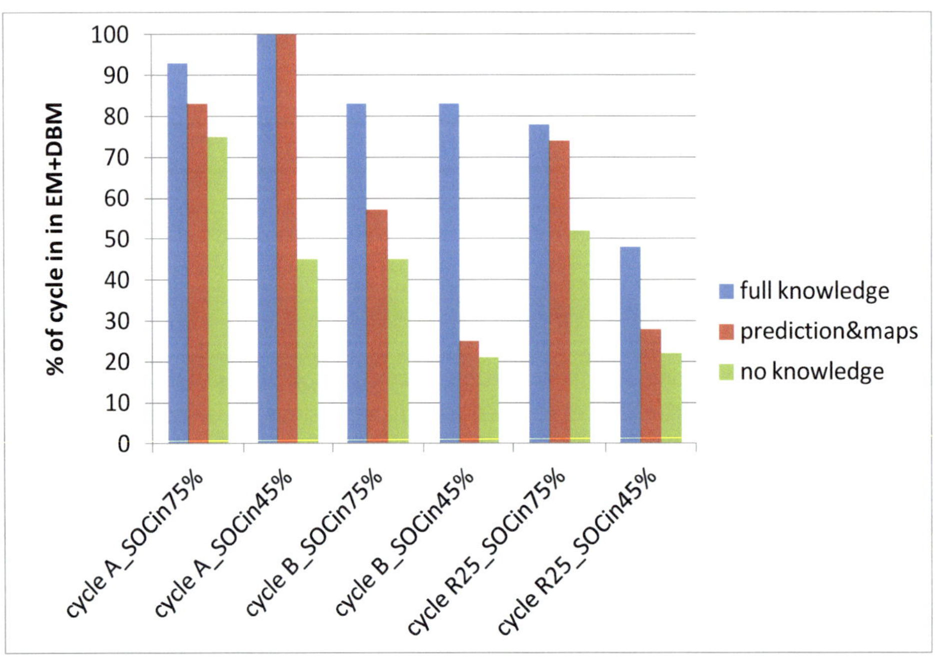

Figure 11. Values of CBD% for cycles A, B and R*25 with SOCin=45% and 75%

By analyzing the results of Figure 16 it is possible to notice that the performance of the proposed strategy is very close to that of full knowledge for cycle A (for both values of SOCin) and for cycle R with SOCin=75%. The values of *CBD%* are always slightly higher than in the *no knowledge* approach for the other cycles.

These results suggest that better performances could be obtained by increasing the duration of the prediction window used in the present investigation (60s) even if this could results in a worse accuracy of the prediction. Future development will be related to the optimization of the prediction horizon to increase the % of the mission covered by EM+DBM.

4.4.2. Percentage of mission with EngineON (engON%)

Another aspect to be taken into account in evaluating the performance of the proposed energy management strategy is the usage of the internal combustion engine in terms of percentage of mission during which the engine is turned ON (*EngON%*).

The results are shown in Figure 12 with respect to cycles A and B to understand the results of Figure 11. Note that cycle B requires the engine to be turned on for a much higher percentage of the mission with respect to cycle A. This explains why this cycle is more critical in the optimization of metric CBD%. Even if the *prediction&maps* is not much successful in optimizing CBD% , it is able to strongly reduce the usage of the engine in both cycle A and B.

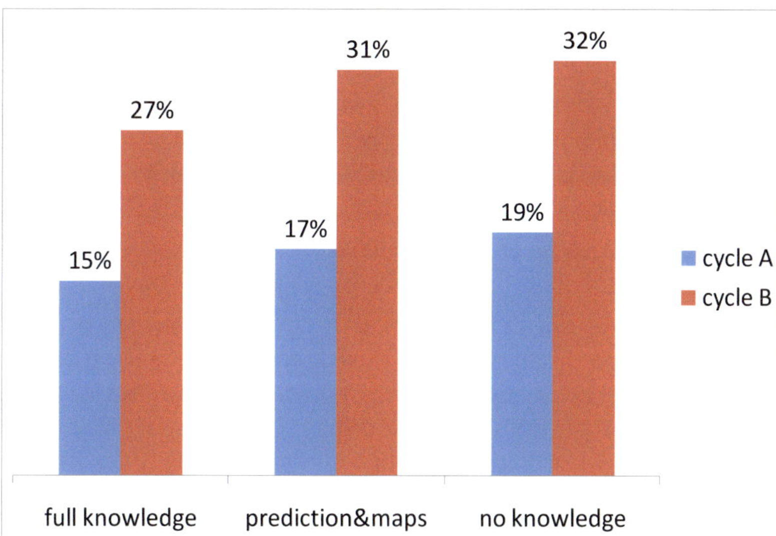

Figure 12. Values of *EngON%*for cycles A and B (SOCin=45%)

4.4.3. AEE (Average Engine Efficiency)

The average efficiency of the engine (AEE) is another important aspect to be taken into account. The results of the comparison are reported in Figure 13.

Once again, the worst performance of the *prediction&maps* method are obtained for cycle B.

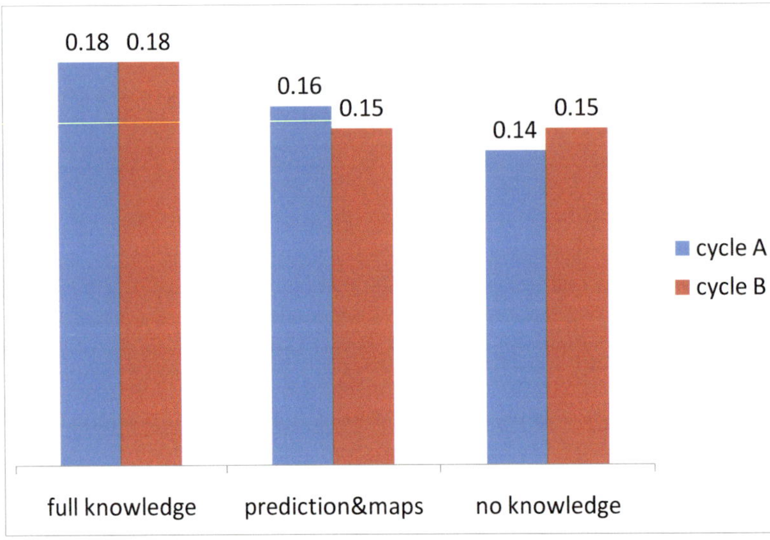

Figure 13. Values of AEE (SOCin=45%)

4.4.4. Well-to-wheel emissions of CO_2

The ultimate goal of advanced power-train technologies is to reduce the overall emissions of greenhouse gases. Thus, it could be interesting to evaluate the overall well-to.-wheel (WTW) emissions of CO_2 produced with the different approaches considered in this investigation.

The complete combustion of 1 liter of gasoline produces 2.4 kg of CO_2. Assuming a density of 700 kg/m3, 1 kg of gasoline produces 3.42 kg of CO_2 (tank to wheel emissions). Sullivan et al. 2004 consider a multiplying factor of 1.162 to pass from TTW to WTW emissions of CO_2. Thus, a kg of gasoline can be assumed to produce 3.98 kg of CO_2 (WTW). Using this conversion factor, the total CO_2 produced along the cycles #A, #B and #R25 has been calculated from the results in Table 3 (i.e. for the full knowledge case).

As for the electric emission, the TTW contribute is obviously zero while the well-to-tank (WTT) emissions depend on the energy mixing used to generate the electricity stored in the batteries. A report from the International Energy Agency, 2011 indicates for Italy an average emission of 0.386 kg of CO_2 per kWh of electric energy. Using the data about the capacity of the batteries (equivalent 1.8 kWh) and the results in terms of SOC, it is possible to evaluate the total energy used for each cycle and for each approach. Thus, the electric WTT emission of CO_2 can be easily calculated.

The calculated values of CO_2 emissions from engine and batteries with the *full-knowledge* approach are reported in Table 6. Note that the electric emissions are almost negligible with respect to the quantity of CO_2 produced by the engine even if the engine is used only for a fraction of the mission. Moreover, they are quite the same for all cycles since the batteries are fully discharged in all cases.

The calculation of the total CO_2 emissions has been repeated for the *no-knowledge* and *prediction&maps* cases. The comparison is shown in Figure 14.

Cycle	Usage of Electric energy	Fuel consuption	CO_2 from engine (WTW)	CO_2 from Battery (WTT)	total CO_2 (WTW)
	[kWh]	[liters]	[kg]	[kg]	[kg]
#A	24.8	1.02	2.92	0.172	3.09
#B	24.7	1.96	5.61	0.172	5.78
#R*25	24.8	1.91	5.47	0.172	5.64

Table 6. Well to wheel emissions of CO_2 in the case of full knowledge

The results of Figure 14 reveal that complete information about the future driving mission could help to significantly reduce the overall emission of CO_2 from a plug-in series HEV. The estimated reduction ranges from 12% for cycle #R*25 to 20% for cycle #A.

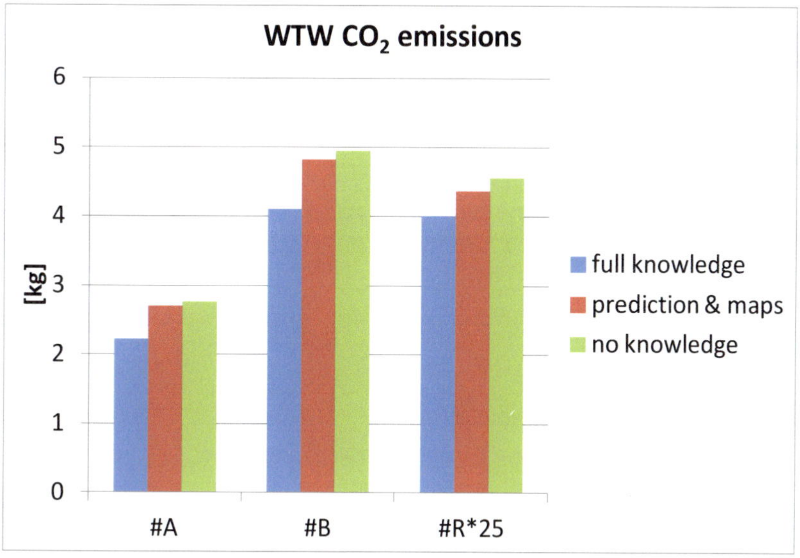

Figure 14. Well to wheel emissions of CO_2 for the proposed approaches

The results of the *prediction&maps* approach are intermediate between full-knowledge and no-knowledge cases. Nevertheless, the results in terms of CO_2 are not satisfactory since the proposed approach helps to reduce the greenhouse emission by only 2-4%. This results suggest the possibility to replace or integrate the goal of the optimization process (eq. 1) with a cost function that takes into account the overall well-to-wheel emission of CO_2. Moreover, better results could be obtained by increasing the duration of the prediction horizon.

5. Summary and conclusions

The chapter describes the optimal usage of an internal combustion engine in an intelligent hybrid electric vehicle able to sense its surrounding and adapt the energy management strategy to the actual driving conditions. After an introduction on hybrid electric vehicles and their challenges, the chapter describes the role of Information and Communication Technologies in the reduction of greenhouse emissions. Then, the chapter focuses on different approaches presented in literature on the usage of information about traffic and weather conditions for the optimal energy management of hybrid electric vehicles. In particular, the chapter describes the application of the *prediction&maps* approach developed at the University of Salento for the optimization of the engine usage in the ITAN500 plug-in hybrid electric vehicle.

Finally, the chapter proposes four metrics to evaluate the performance of the proposed method: the percentage of mission performed before reaching the lowest allowed value for battery state of charge (CBD%), the percentage of mission execute with the engine turned ON (EngON%), the average efficiency of the engine (AEE), calculated according to its actual temperature and the overall well-to-wheel emissions of CO_2.

Author details

Teresa Donateo
University of Salento, Italy

Acknowledgement

The investigation was supported by the University of Salento and the Italian Ministry for Environment (MATTM), through the funding of the "P.R.I.M.E." project.

List of acronyms

AEE	All Electric Range
BEV	Battery Electric Vehicles
CBD%	% of mission with controlled battery discharge
CD	Charge Depleting
CS	Charge Sustaining
ECMS	Equivalent Consumption Minimization Strategy
EngON%	% of mission with engine turned on
FCV	Fuel-Cell Vehicles
GPS	Global Positioning System
HEV	Hybrid Electric Vehicles
PHEV	Plug-in Hybrid Electric Vehicles
ICE	Internal Combustion Engine
SOC	State of Charge
TTW	Tank-to-Wheel
WTT	Well-to-Tank
WTW	Well-to-Wheel

6. References

Anatone M., Cipollone R., Sciarretta A. (2011), "Control-Oriented Modeling and Fuel Optimal control of a Series Hybrid Bus", SAE paper 2005-01-1163.

Barsali S., Miulli C., Possenti A. (2004), "A Control Strategy to Minimize Fuel Consumption of Series Hybrid Electric Vehicles", *IEEE Transactions of Energy Conversion*, Vol. 19, no. 1, pp. 187-195.

Bayar K., Bezaire B., Cooley B., Kruckenberg J., Schacht E., Midlam-Mohler S., and Rizzoni G. (2010), "Design of an extended-range electric vehicle for the EcoCAR challenge," *ASME Conference Proceedings*, vol. 2010, no. 44090, pp. 687–700.

Chan C.C. (2007), "The state of the Art of Electric, Hybrid, and Fuel Cell Vehicles", *Proceeding of the IEEE*, Vol. 95, No.4.

Ciccarese G., Donateo T., Palazzo C. (2012 in printing), "On-Board Prediction of Future Driving Profile for the Sustainable Mobility", *Int. J. Automotive Technology and Management*, Vol. x, No. x, ISSN (Online): 1741-5012 - ISSN (Print): 1470-9511.

Ciccarese G., Donateo T.,Marra P., Pacella D., Palazzo C. (2010), "On the Use of Vehicular Communications for Efficient Energy Management of Hybrid Electric Vehicles, *Proceeding of Fisita 2010*, Paper n. F2010E047.

Dimitrakopoulos G. (2011), "Intelligent Transportation Systems based on Internet-Connected Vehicles: Fundamental Research Areas and Challenges", 11th International Conference on ITS Telecommunications.

Donateo T., Pacella D. (2012), "Modeling the Thermal Behavior of Internal Combustion in Hybrid Electric Vehicles with and without Exhaust Gas Heat Recirculation", *Proceedings of the ASME 2012 Internal Combustion Engine Division Spring Technical Conference*, May 6-9, 2012.

Donateo T., Pacella D., Laforgia D. (2011), "Development of an Energy Management Strategy for Plug-in Series Hybrid Electric Vehicle Based on the Prediction of the Future Driving Cycles by ICT Technologies and Optimized Maps", SAE Technical Paper 2011-01-0892.

Ericsson E. (2001), "Independent driving pattern factors and their influence on fuel-use and exhaust emission factors", *Trans. Res.*, Part D, vol. 6, no. 5, pp. 325–341.

Ericsson E. (2000), "Variability in urban driving patterns", *Trans. Res.*, Part D, vol. 5, no. 5, pp. 337–354.

German J.M. (2003), *Hybrid Powered Vehicles*, SAE International, ISBN 0-7680-1310-0.

Gonder J. and Markel T. (2007), "Energy Management Strategies for Plug-In Hybrid Electric Vehicles", SAE Technical paper 2007-01-0290.

Gong Q., Tulpule P., Marano V., Midlam-Mohler S., Rizzoni G. (2011), "The role of ITS in PHEV Performance Improvement", American Control Conference ACC.

Gong Q., Li Y., Peng R. (2007), "Optimal Power Management of plug-in HEV with intelligent transportation systems", IEEE/ASME International Conference on Advanced Intelligent Mechantronics.

Gusikhin O., Filev D., Rychtyckyj N. (2008), "Intelligent Vehicle Systems: Applications and New Trends", *Lecture Notes in Electrical Engineering*, Volume 15, Part 1, 3-14.

Guzzella L., Onder C.H. (2004), *Introduction to Modeling and Control of Internal Combustion Engine Systems*, Springer Verlag, ISBN 354022274X, 2nd edition.

Guzzella L., Sciarretta, A. (2007), *Vehicle Propulsion Systems, Introduction to Modelling and* Optimization, Springer Verlag, ISBN 3540746919, 2nd edition.

International Energy Agency (2011), "CO_2 Emissions from Fuel Combustion - 2011 Report".

Janota A., Dado M., Spalek J. (2010), "Greening Dimension of Intelligent Transport", *Journal of Green Energy*, Vol 1-1

Karbowski D., Rousseau A., Pagerit S., Sharer P. (2006), "Plug-in Vehicle Control Strategy: from Global Optimization to real-time application", 22nd Electric Vehicle Symposium, EVS22, Yokohama, Japan.

Lee J., Ohn H., Choi J-Y., Kim S. J., Min B. (2011) "Development of Effective Exhaust Gas Heat Recovery System for a Hybrid Electric Vehicle", SAE Technical Paper 2011-01-1171;

Lin C.-C, Peng H., Grizzle J. W., and Kang J.-M. (2003), "Power management strategy for a parallel hybrid electric truck", *IEEE Trans. Control Syst. Technol.*, vol. 11, no. 6, pp. 839–849.

Markel T., Kuss M., Denholm P. (2009), "Communication and Control of Electric Drive Vehicles Supporting Renewables", Vehicle Power and Propulsion Conference, 2009. VPPC '09. IEEE.

Millo F., Rolando L., Servetto E. (2011), "Development of a Control Strategy for Complex Light-Duty Diesel-Hybrid Powertrains", SAE Technical paper 2011-24-0076.

Paladini V., Donateo T., de Risi A., Laforgia D. (2007), "Super-capacitors fuel-cell hybrid electric vehicle optimization and control strategy development", *Energy Conversion and Management*, 48 (11), p.3001-3008, ISSN: 0196-8904.

Schuricht P., Cassebaum O., Luft M., Baker B. (2010), "Methods and Algorithms in Control of Hybrid Powertraines, International congress of heavy vehicles, road trains and urban transport, 06-09 October 2010.

Sciarretta A., Guzzella L., and Back M. (2004), "A real-time optimal control strategy for parallel hybrid vehicles with on-board estimation of the control parameters," in *Proc. IFAC Symp. Adv. Autom. Control*, Salerno, Italy, April 19–23, 2004.

Serrao L, "A Comparative Analysis of Energy Management Strategies for Hybrid Electric Vehicles" (2009), Ph.D. Dissertation, The Ohio State University.

Serrao L., Rizzoni G. (2008), "Optimal Control of Power Split for a Hybrid Electric Refuse Vehicle", *Proceedings of the 2008 American Control Conference*.

Simpson A. (2006), "Cost-benefit Analysis of Plug-In Hybrid Electric Vehicle Technology", 22nd International Battery, Hybrid and Fuel Cell Electric Vehicle Symposium and Exhibition, 23- 28 October,Yokohama, Japan;

Sullivan, J.L., Baker, R.E., Boyer B.A., Hammerle R.H., Kenney T.E., Muniz L., Wallington T.J. (2004), "CO_2 Emission Benefit of Diesel (versus Gasoline) Powered Vehicles", *Environmental Science & Technology*, Vol. 38 No. 12.

Tulpule P., Marano V., Rizzoni G. (2009), "Effects of Different PHEV Control Strategy on Vehicle Perfomance", American Control Conference DSCC 2009.

Tulpule, P., Marano, V., and Rizzoni, G. (2011), "Effect of Traffic, Road and Weather Information on PHEV Energy Management," SAE Technical Paper 2011-24-0162.

Weng Y., Wu T. (2011), "Car-following model of vehicular traffic", International Conferences on Info-tech and Info-net 2001, ICII 2001, Oct. 2001.

Won J.-S. and Langari R. (2005), "Intelligent energy management agent for a parallel hybrid vehicle—Part 2: Torque distribution, charge sustenance strategies, and performance results", *IEEE Trans. Veh. Technol.*, vol. 54, no. 3, pp. 935–953.

Yu S., Li L., Dong G. and Zhang X. (2006), "A Study of Control Strategies of PFI Engine during Cranking and Start for HEVs, *Proceedings of IEEE International Conference on Vehicular Electronics and Safety*, ICVES 2006.

Zuurendonk, B. (2005), Advanced Fuel Consumption and Emission Modeling using Willans line scaling techniques for engines", traineeship report, DCT 2005.116, Technische Universiteit Eindhoven.

Chapter 7

Modeling and Simulation of SI Engines for Fault Detection

Mudassar Abbas Rizvi, Qarab Raza,
Aamer Iqbal Bhatti, Sajjad Zaidi and Mansoor Khan

Additional information is available at the end of the chapter

http://dx.doi.org/10.5772/50487

1. Introduction

During last decades of twentieth century, the basic point of concern in the development of Spark Ignition engine was the improvement in fuel economy and reduced exhaust emission. With tremendous of electronics and computer techniques it became possible to implement the complex control algorithms within a small rugged *Electronic Control Unit* (ECU) of a vehicle that are responsible to ensure the desired performance objectives. In modern vehicles, a complete control loop is present in which throttle acts as a user input to control the speed of vehicle. The throttle input acts as a manipulating variable to change the *set point* for speed. A number of sensors like *Manifold Air Pressure* (MAP), Crankshaft Speed Sensor, Oxygen sensor etc are installed in vehicle to measure different vehicle variable. A number of controllers are implemented in ECU to ensure all the desired performance objectives of vehicle. The controllers are usually designed on the basis of mathematical representation of systems. The design of controller for SI engine to ensure its different performance objectives needs mathematical model of SI engine. Mean Value Model (MVM) is one of the most important mathematical models used most frequently by the research community for the design of controllers; see for example [1], [2], [5], [7], [9], [13], [14], and [15]. The basic mean value model is based on the average behavior of SI engine in multiple ignition cycles.

Although the controllers implemented in vehicle ECU are sufficiently robust, yet introduction of fault in system significantly deteriorate the system performance. Research is now shifted to ensure the achievement of performance objectives even in case of some fault. The automotive industry has implemented some simple fault detection algorithms in ECU that identify the faults and provide their indication to a fault diagnostic kit in the form of some fault codes. The implementation is however crude as it provides fault indication only when the fault become significant. For incipient faults, the vehicle would keep its operation but under sub-optimal conditions till the magnitude of fault would grow to such an extent that it would become visible. Again mathematical models are used to identify the faults and develop techniques to detect the engine faults.

Different mathematical models of Spark Ignition engine proposed in literature in recent years along with the domain of their application are reviewed. The emphasis would however be given to two different mathematical models

- The version of mean value model proposed by the authors [11], [12].
- Hybrid model proposed by the authors [21], [22].

The first section of this chapter would present different models of SI engine with only a brief description of those models. The second section would give the mathematical development of mean value model along with the simulation results of presented model and its experimental validation. The third section would give the mathematical derivation of Hybrid model along with the simulation results of model and experimental verification of simulation results. The fourth section would identify the application of these models for fault diagnosis applications.

2. Review of models of spark ignition engine

The dynamic model of a physical system consists of a set of differential equations or difference equations that are developed under certain assumptions. These mathematical models represent the system with fair degree of accuracy. The main problem in development of these models is to ensure the appropriateness of modeling assumptions and to find the value of parameters that appear in those equations. However the basic advantage of this approach is that the develop model would be generic and could be applied to all systems working on that principle. Also the model parameters are associated with some physical entity that provides better reasoning. An alternate modeling technique is to represent the system using neural network that is considered to be a universal estimator. A suitably trained neural network sometime represents the system with even better degree of accuracy. The main problems associated with this approach are the lack of any physical reasoning of parameters, appropriate training of neural network and lack of generality i.e. a neural network model trained on one setup may not work properly on another setup of similar nature.

The research community working on mathematical modeling of SI engine has used both these approaches for control, state estimation and diagnostic applications. In this regard a number of different models were developed to represent the SI engine using both these approaches. Mean Value Model (MVM), Discrete Event Model (DEM), Cylinder by Cylinder Model, Hybrid Model are some of the key examples of models developed using basic laws of physics. Most neural network based models are based in one way or other on Least Square Method. A brief description of some of these models is below.

2.1. Mean Value Model (MVM)

Mean Value Model is developed on the basis of physical principles. In this model throttle position is taken as input and crankshaft speed is considered to be the output. A careful analysis indicates that MVM proposed by different researchers share same physical principles but differ from each other slightly in one way or the other [1-16], [26], and [27]. The idea behind the development of model is that the output of model represents the average response of multiple ignition cycles of an SI engine although the model could be used for cycle by cycle analysis of engine behavior. The details about the development of MVM on the basis of physical principles are provided in section 3 of this chapter.

2.2. Discrete Event Model (DEM)

An SI engine work on the basis of Otto cycle in which four different processes i.e. suction, compression, expansion and exhaust take place one after the other. In a four stroke SI engine, each of these processes occurs during half revolution (180°) of engine shaft. Therefore irrespective of the engine speed it always takes two complete rotations of engine shaft to complete one engine cycle. The starting position of each of the four processes occurs at fixed crank position but depend upon certain events e.g. expansion is dependent on spark that occur slightly ahead of Top Dead Center (TDC) of engine cylinder. Also with Exhaust Gas Recycling (EGR), a portion of exhaust gases are recycled in suction. Due to EGR some delay is present in injection system to ensure overlap between openings of intake valve and closing of exhaust valve.

The working of SI engine indicates that the link of engine processes is defined accurately with crankshaft position. In discrete engine model, crankshaft position is taken as independent variable instead of time. Mathematical model based on the laws of physics is developed for air flow dynamics and fuel flow dynamics in suction and exhaust stroke, production of torque during power stroke. The crankshaft speed is estimated by solving the set of differential equations of all these processes for each cylinder. Computational cost of DEM is high but it can identify the behavior of engine within one engine cycle. Modeling the discreet event model could be seen in [1].

2.3. Cylinder by Cylinder Model (CCM)

In these models, the forces acting on piston of each cylinder are modeled on the basis of laws of physics. The input to these models is the forces acting on the crankshaft assembly and output is the crankshaft speed. The forces acting on crankshaft assembly are estimated using pressure established inside the cylinder due to the burning of air fuel mixture. For a comparison of MVM and CCM, see [23].

2.4. Hybrid model

Hybrid model represent the integration of continuous dynamics and discrete events in a physical system [19], [21], and [22]. In SI engine, the variables like crankshaft speed represent continuous dynamics but the spark is a discrete event. In hybrid model, the four cylinders are considered four independent subsystems and are modeled as continuous system. The cylinder in which power stroke occur is considered as the active cylinder that define the crankshaft dynamics. The sequence of occurrence of power stroke in four cylinders is defined as a series of discrete events. The behavior of SI engine is defined by the combination of both of them. The details of hybrid dynamics is provided in section 4 of this chapter.

3. Mean Value Model (MVM)

In this section a simple nonlinear dynamic mathematical model of automotive gasoline engine is derived. The model is physical principle based and phenomenological in nature. Engine dynamics modeled are inlet air path, and rotational dynamics. A model can be defined as

"*A model is a simplified representation of a system intended to enhance our ability to understand, explain, change, preserve, predict and possibly, control the behavior of a system*"[25].

When modeling a system there are two kinds of objects taken into consideration

- Reservoirs of energy, mass, pressure and information etc
- Flows of energy, mass, pressure and information etc flowing between reservoirs due to the difference of levels of reservoirs.

An MVM should contain relevant reservoirs only but there are no systematic rules to decide which reservoirs to include in what model. Only experience and iterative efforts can produce a good model. The studied machine is a naturally breathing four-stroke gasoline engine of a production vehicle equipped with an ECU compliant to OBD-II standard. The goal is to develop a simple system level model suitable for improvement of model-based controller design, fault detection and isolation schemes. The model developed in here has following novel features.

- Otto (Isochoric) cycle is used for approximation of heat addition by fuel combustion process.
- Consequently the maximum pressure inside the cylinder and mean effective pressure (MEP) are computed using equations of Otto Cycle for prediction of indicated torque. A detailed description of Otto cycle is available in most thermodynamics and automotive engine text books.
- Fitting/ regressed equations based on experimental data and constants are avoided except only for model of frictional/pumping torque which has been adapted from available public literature [26], and [27] and modified a little bit.

The model is verified with data obtained from a production vehicle engine equipped with an ECU compliant to OBD-II. Most of the models available in literature are specific to a certain brand or make because of their use of curve fittings, thus limiting their general use. Here a model is proposed which is not confined to a certain engine model and make; rather it is generic in nature. It is also adaptable to any make and model of gasoline engine without major modifications. Following are outlines of framework for deriving this model.

1. It is assumed that engine is a four stroke four cylinder gasoline engine in which each cylinder process is repeated after two revolutions.
2. It is also assumed that cylinders are paired in two so that pistons of two cylinders move simultaneously around TDC and BDC but only one cylinder is fired at a time. Due to this, one of the four principle processes namely suction, compression, power generation and exhaust strokes, is always taking place in any one of cylinders at a time. Therefore the abovementioned four engine processes can be comfortably taken as consecutive and continuous over time. This assumption is due to the fact that each one of the four processes takes a theoretical angular distance of π radians to complete and hence in a four stroke engine considered above one instance of each of the four processes is always taking place in one of the cylinders.
3. The fluctuations during power generation because of gradual decrease in pressure inside the cylinder during gas expansion process (in power strokes) are neglected and averaged by mean effective pressure (MEP) which is computed using Otto cycle as mentioned earlier. This simplifies the model behavior maintaining the total power output represented by the model. The instantaneous combustion processing modeling and consequent power generation model is complex and require information about cylinder inside pressure and temperature variations at the time of spark and throughout power stroke. Moreover, the combustion and flame prorogation dynamics are very fast and usually inaccessible for a controller design perspective.
4. The fluctuations of manifold pressures due to periodic phenomena have also been neglected. Equations of manifold pressure and rotational dynamics have been derived using the physics based principles.
5. The exhaust gas recirculation has been neglected for simplicity.
6. The choked flow conditions across the butterfly valve have also been neglected because in the opinion of author, sonic flow rarely can occur in natural breathing automotive gasoline engines due to the nonlinear coupling and dependence of air flow and manifold pressure on angular velocity of crankshaft.
7. It is also assumed that the temperature of the manifold remains unchanged for small intervals of time; therefore the manifold temperature dynamics have been neglected at this time and it is taken to be a constant.
8. Equation representing the rotational dynamics has been developed using Newton's second law of motion. Otto cycle has been used for combustion process modeling, hence the computation of maximum cycle pressure, maximum cycle temperature, Mean Effective Pressure (MEP), and indicated torque (T_i).
9. The equations of frictional and pumping torque have been taken from available public literature, because frictional torque is extremely complex quantity to model for an IC engine due to many a number of rotating and sliding parts, made of different materials, changing properties with wear/tear and aging, and variations of frictional coefficients of these parts, changing properties of lubricating oil on daily basis.

The air dynamics are further divided into throttle flow dynamics, manifold dynamics and induction of air into the engine cylinders. These are separately treated below and then combined systematically to represent the induction manifold dynamics.

3.1. Throttle flow dynamics

Throttle flow model predicts the air flowing across the butterfly valve of throttle body. The throttle valve open area has been modeled by relationships of different levels of complexities for accuracy, see for example [2] and [16], but here it is modeled by a very simple relationship as

$$A(\alpha) = (1 - \cos\alpha)\frac{\pi}{4}D_T^2\ ,\ \alpha_0 \leq \alpha \leq \alpha_{max} \tag{1}$$

Where $\frac{\pi}{4}D_T^2$ is cross sectional area of throttle valve plate with D_T, being the diameter of plate facing the maximum opening of pipe cross section, α is the angle at which the valve is open and $A(\alpha)$ is the effective open area for air to pass at plate opening angle α. The angle α_0 is the minimum opening angle of throttle plate required to keep the engine running at a lowest speed called idle speed. At this point engine is said to be idling. The angle α_{max} is the maximum opening angle of throttle plate, which is 90°. The anomalies arising will be absorbed into the discharge coefficient($\boldsymbol{C_d}$).

Mass flow rate across this throttle valve ($\dot{m}_{ai}$) is modeled with the isentropic steady sate energy flow equation of gases and the derived expression is as below.

$$\dot{m}_{ai} = A(\alpha)P_a\boldsymbol{C_d}\sqrt{\frac{2}{\beta RT_a}}\sqrt{\left(\left(\frac{P_m}{P_a}\right)^{\zeta} - \left(\frac{P_m}{P_a}\right)^{\xi}\right)},\ \frac{P_m}{P_a} < 1 \tag{2}$$

$$\text{where } \beta = \frac{\gamma - 1}{\gamma}, \zeta = \frac{2}{\gamma} \text{ and } \xi = \frac{\gamma + 1}{\gamma} \tag{3}$$

Here $A(\alpha)$ is defined in equation (1), P_a is atmospheric pressure, P_m is intake manifold pressure, R is universal gas constant, T_a is ambient temperature, and γ is specific heat ratio for ambient air.

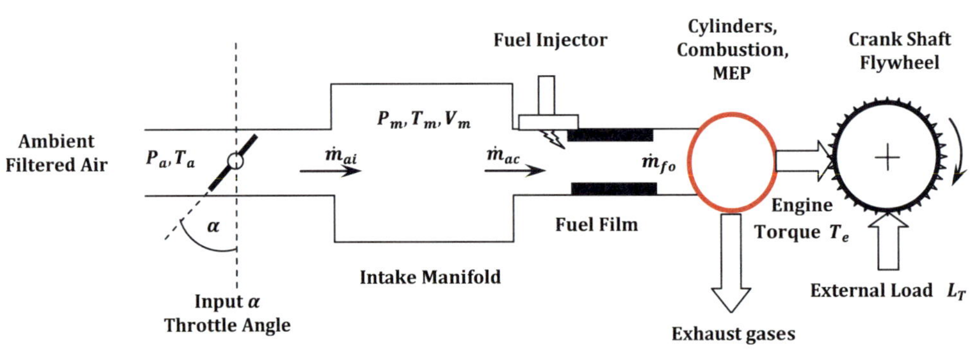

Figure 1. Diagram showing the components of a Mean Value Model of Gasoline Engine

3.2. Air induced in cylinders

The air mass induced into the cylinder ($\dot{m}_{ac}$) is modeled with speed density equation of reciprocating air pumps/ compressors, because during suction stroke the engine acts like one. The expression for an ideal air pump is given by the equation:

$$\dot{m}_{ac} = \rho V_d N \tag{4}$$

Where ρ is the density of air, V_d is swept volume of engine cylinders, and N is crankshaft speed in rev/min (rpm). In terms of variables easily accessible for measurement, the expression can be converted into the following using $\rho = \frac{P_m}{RT_m}$ and $V_d N = C_0 \omega$. Here N is in rpm and ω in rad/s and C_0 holds all the necessary conversions.

$$\dot{m}_{ac} = \frac{P_m}{RT_m} C_0 \omega \tag{5}$$

Since air compresses and expands under varying conditions of temperature and pressure, therefore the actual air induced into the cylinder is not always as given by the equation. Hence an efficiency parameter called volumetric efficiency (η_v) is introduced which determines how much air goes into the engine cylinder. The equation therefore can be written as below:

$$\dot{m}_{ac} = \frac{P_m}{RT_m} \eta_v C_0 \omega \tag{6}$$

3.3. Intake manifold dynamics

The intake manifold dynamics are modeled with filling and emptying of air in the intake manifold. The manifold pressure dynamics are created by filling of inlet manifold by mass flow of air entering from the throttle valve ($\dot{m}_{ai}$) and emptying of the manifold by expulsion of air and flow into the engine cylinder($\dot{m}_{ac}$). Using ideal gas equation for intake manifold this can be derived as

$$\begin{aligned} &PV = mRT \\ &P_m V_m = m_m R T_m \quad \text{(Using the relationship in manifold variables)} \\ &\Rightarrow P_m = \frac{RT_m}{V_m} m_m \end{aligned} \tag{7}$$

Here *P*, *V*, *m*, *R* and *T* are pressure, volume, mass, Gas constant and temperature of air. It was assumed that the manifold temperature variations are small, and therefore manifold temperature is taken to be constant. To this reason, its differentiation is neglected and only variables are taken to be mass flow and manifold pressure.

$$\Rightarrow \dot{P}_m = \frac{RT_m}{V_m} \dot{m}_m \quad \text{(Differentiation w.r.t. time)} \tag{8}$$

The quantity $\dot{m}_m$ represents the instantaneous mass variation from filling and emptying of intake manifold, assuming $\dot{m}_m = \dot{m}_{ai} - \dot{m}_{ac}$ we can write it as

$$\dot{P}_m = \frac{RT_m}{V_m}(\dot{m}_{ai} - \dot{m}_{ac}) \tag{9}$$

Putting (2) and (4) in (7), and simplifying the resulting equation gives us the required equation of manifold dynamics.

$$\dot{P}_m = \frac{RT_m}{V_m}\dot{m}_{ai} - \frac{RT_m}{V_m}\frac{P_m}{RT_m}\eta_V C_0 \omega$$

$$\Rightarrow \dot{P}_m = \frac{RT_m}{V_m}\dot{m}_{ai} - C_1 \eta_v P_m \omega \tag{10}$$

$$\text{where } C_1 = \frac{1}{V_m} C_0$$

Putting the expression of air mass flow from (2), this equation can also be written in the following form.

$$\dot{P}_m = \frac{RT_m}{V_m} A(\alpha) P_a C_d \sqrt{\frac{2}{\beta R T_a}} \sqrt{\left(\left(\frac{P_m}{P_a}\right)^{\zeta} - \left(\frac{P_m}{P_a}\right)^{\xi}\right)} - C_1 \eta_v P_m \omega \tag{11}$$

Where $\beta, \zeta,$ and ξ have already been defined in (2A).

3.4. Rotational dynamics

Rotational dynamics of engine are modeled using mechanics principles of angular motion. Thus torque T_b produced at the output shaft (also called brake torque) of the engine is given by Newton's second law of motion as

$$T_b = J_e \alpha \quad \text{Or} \tag{12}$$

Where J_e is the rotational moment of inertia of engine rotating parts and α is angular acceleration. This can also be represented as given below.

$$\alpha = \frac{1}{J_e} T_b \text{ Or } \dot{\omega} = \frac{1}{J_e} T_b \tag{13}$$

The above relationship represents the rotational dynamics in general form. The brake torque is a complex quantity and is a sum of other torque quantities, which are indicated torque T_i, frictional torque T_f, pumping torque T_p, and load torque L_T, the external load on engine. Indicated torque comes from the burning of fuel inside the cylinder. Frictional torque is the power loss in overcoming the friction of all the moving parts (sliding and rotating) of the engine, for example, piston rings, cams, bearings of camshaft, connecting rod, crankshaft etc to name a few. Pumping torque represents the work done by engine during the compression of air and consequently raising the pressure and temperature of fresh air and fuel mixture trapped inside the cylinder during compression stroke. Load torque is work done by engine in running/pulling of vehicle, its passengers, goods and all the accessories. These are briefly described in 3.4.3. Mathematically brake torque is given by the following relationship.

$$T_b = T_i - T_p - T_f - L_T \tag{14}$$

Sometimes the quantities other than L_T in (14) are called Engine Torque $T_e = T_i - T_p - T_f$ and above equation is written as

$$T_b = T_e - L_T \tag{15}$$

In this work the form given in (14) and not in (15), will be maintained for parameterization purpose. The Torque quantities in (12) are defined as below.

3.4.1. Indicated torque ($\boldsymbol{T_i}$)

Indicated torque T_i is the theoretical torque of a reciprocating engine if it is completely frictionless in converting the energy of high pressure expanding gases inside the cylinder into rotational energy. Indicated quantities like indicated horsepower, and indicated mean effective pressure etc. are calculated from indicator diagrams generated by engine indicating devices. Usually these devices consist of three basic components which are:

- A pressure sensor to measure the pressure inside engine cylinder.
- A device for sensing the angular position of crankshaft or piston position over one complete cycle.
- A display which can show the pressure inside cylinder and volume displaced on same time scale.

For a physics based engine model, indicated torque has to be estimated through any of the various estimation techniques. In this work indicated torque is presented as a function of manifold pressure. To do this, the engine processes are modeled using Otto Cycle. For computation of Indicated Torque another quantity, Mean Effective Pressure (MEP) is computed first which is given by the relationship below.

$$MEP = \frac{{C_r}^{2-\gamma}(C_r^{\gamma-1}-1)(H_k Q)}{(\gamma-1)(C_r-1)C_v T_m (AFR)} P_m \tag{16}$$

Here C_r is the compression ratio of engine, Q is calorific value of fuel (gasoline etc.), H_k is fraction of burnt fuel heat energy available for conversion into useful work, C_v is specific heat of air at constant volume, T_m is intake manifold temperature, and AFR is the stoichiometric air to fuel ratio. Mean Effective Pressure is defined as the average constant pressure which acts on the piston head throughout the power stroke (pressure that remains constant from TDC to BDC). In actual practice, the high pressure generated by combustion starts decreasing as the piston moves away from TDC and burnt gases start expanding. Since the mean effective pressure is computed on the basis of Otto cycle; the thermal efficiency ($\eta_{th} = 1 - \frac{1}{C_r^{\gamma-1}}$) of Otto cycle must be considered when calculating indicated torque. Indicated torque is given as the product of MEP, η_{th}, and volume displaced per second.

$$T_i = \frac{V_d}{4\pi}\eta_{th} MEP = \frac{V_d}{4\pi}\left(\frac{{C_r}^{2-\gamma}(C_r^{\gamma-1}-1)(H_k Q)\left(1-\frac{1}{C_r^{\gamma-1}}\right)}{(\gamma-1)(C_r-1)C_v T_m (AFR)}\right) P_m \tag{17}$$

The state variable in the above expression is manifold pressure (P_m). All the other quantities/ parameters are constants or taken to be constant usually. For example, the displacement volume of engine under consideration (V_d) is a strictly constant value. The same is true for compression ratio (C_r); this variable is a particular number for a production vehicle. For example, this number is 8.8 ($C_r = 8.8$) for engine under study. The calorific value of fuel (Q), manifold temperature (T_m), and H_k are also taken to be fixed constants. As long as other parameters of the above expression are concerned, with the ambient air and atmospheric conditions of pressure and temperature, the variation in their values is very low but inside the cylinder, and during and after combustion, not only the composition of air changes, but also its properties may vary. The variation of index of expansion (γ) with density and composition of a gas is well documented in public literature. To accommodate all the variations and some heat transfer anomalies, the entire expression is written as following, making indicated torque a function of manifold pressure with a time varying parameter (a_1). This parameter is called indicated torque parameter and will be estimated later in this work. In a proper way this parameter may be written as $a_1(t)$ but the brackets and variable t are omitted for simplicity. With all this, the expression of indicated torque in (17) becomes

$$T_i = a_1 P_m \tag{18}$$

3.4.2. *Frictional and pumping torque* ($\boldsymbol{T_f},\ \boldsymbol{T_p}$)

The Modeling of frictional and pumping torque has been done using a well known empirical relationship given by the equation as

$$T_f = \frac{1}{2\pi} V_d \{97000 + 15N + 5N^2 10^{-3}\} = b_1 + a_2\omega + a_3\omega^2 \tag{19}$$

A slight variation of that can be found in [26] and [27]. The equation has been converted into the variable ω with necessary conversion factors. Also the constant b_1 is merged into the load torque L_T. Therefore the minimum value of load torque is equal to or greater than b_1 even when engine is idling. We can write it as

$$L_T = b_1 + \text{External Load on Engine}$$

The relationship given in (19) represents the quantity T_f for throttle positions closer to WOT (wide open throttle) and for engines up to 2000 cc [26].

3.4.3. *Load torque* ($\boldsymbol{L_T}$)

Load torque L_T is external load on engine. It is the load the engine has to pull/ rotate, and it includes all other than the frictional losses all around and pumping work. In case of a vehicle, all the rotating parts of engine and its driven subsystems, including electrical generating set, cam and valve timing system, air conditioner etc. and beyond the clutches, toward the differential gear assembly and wheels, the weight of vehicle and everything in it is the load while in case of electrical generating set, the generator is the load. In a production vehicle, a significant part of the load is also created at random due to driver commands and road/path conditions; for example, the climbing road poses a greater load as compared to the flat roads (without climbing slope) because of the torque required to work against gravity. Conversely on a downward slope, the vehicle may have an aid in moving down due to roller coaster effect. In urban areas, with random turnings, road lights, and random traffic etc. the load on engine cannot be predicted *apriori*, and has to be estimated.

3.5. Fuel dynamics

The fuel dynamics are considered to be ideal. Since fuel dynamics also have parametric variations and the delays due to the internal model feedback loops; taking ideal fuel dynamics will ensure that the model simulations are free from interferences of fuel dynamics parametric noise. The model of fuel flows is given here for completeness only [9].

$$\dot{m}_{fc} = \dot{m}_{ff2} + \dot{m}_{ff3} + \dot{m}_{fsl} \tag{20}$$

The components of this model are:

- $\dot{m}_{fc}$ Mass of fuel entering into the engine cylinder in air fuel mixture.
- $\dot{m}_{ff2}$ Mass of fuel from injector before inlet valve closes.
- $\dot{m}_{ff3}$ Mass of fuel from injector after inlet valve closes, and entering in cylinder in next engine cycle.
- $\dot{m}_{fsl}$ Mass of fuel lagged due to liquid film formation and re-evaporation.

3.6. Model summary

The model derived and described in previous sections is a three state nonlinear model. It can be represented as a set of dynamical equations given below:

Fuel flow dynamics

$$\dot{m}_{fc} = \dot{m}_{ff2} + \dot{m}_{ff3} + \dot{m}_{fsl} \tag{21}$$

Manifold dynamics

$$\dot{P}_m = \frac{RT_m}{V_m} A(\alpha) P_a C_d \sqrt{\frac{2}{\beta R T_a}} \sqrt{\left(\left(\frac{P_m}{P_a}\right)^{\zeta} - \left(\frac{P_m}{P_a}\right)^{\xi}\right)} - C_1 \eta_v P_m \omega + \tag{22}$$

Rotational dynamics

$$\dot{\omega} = a_1 P_m - a_2 \omega - a_3 \omega^2 - L_T \tag{23}$$

Where

$$a_1 = \frac{1}{J_e} \frac{V_d}{4\pi} \left(\frac{{C_r}^{2-\gamma} (C_r^{\gamma-1} - 1)(H_k Q)\left(1 - \frac{1}{C_r^{\gamma-1}}\right)}{(\gamma-1)(C_r-1) C_v T_m (AFR)} \right) 10^3 \tag{24}$$

$$a_2 = \frac{1}{J_e}\frac{V_d}{4} \tag{25}$$

$$a_3 = \frac{1}{J_e}\frac{V_d.(0.05\pi)}{18\times 10^4} \tag{26}$$

A detailed description of nonlinear engine models and their background can be studied in [1-16]. With fuel dynamics considered to be ideal, the model becomes a two state nonlinear model consisting of (17) and (18) only.

3.7. Model simulations and engine measurements

There is a great difference between theory and practice.

Giacomo Antonelli (1806-1876)

The manifold dynamics equation derived in earlier section is simulated on a digital computer. The simulation software used is Matlab and Matlab Simulink©. The S-function template available in Matlab is used to program the dynamic model and graphical interface of Simulink© is to run the simulations. The engine measurements are taken using an OBD-II compliant scanning hardware and windows based scanning and data storing software. The model is primed with the same input as engine was and manifold pressure measurements and model manifold pressures are plotted and compared.

A couple of set of simulations are presented for two values of discharge coefficient, and the patterns of engine measurement of manifold pressure and the output of manifold dynamic equation are compared. In first set of simulation, the discharge coefficient is taken to have its ideal value which is 1.0; and the model output manifold pressure is compared with engine measured manifold pressure. While in another simulation test, the discharge coefficient is taken to be equal to 0.5 and the experiment is repeated. As we can see from the Figure 4 and Figure 5 that the shape of trajectory of model manifold pressure and engine measurements is a large distance apart. Moreover, these trajectories do not follow the same shape and pattern. From which we can comfortably deduce that both trajectories cannot be made identical by scaling with a constant number only; thus discharge coefficient of the derived model should not be a constant number. Also, at certain points in time, the evolution of both trajectories is opposite in directions. From all of this it can be concluded that the discharge coefficient should be considered a time varying parameter.

The above figure shows the first simulation of model derived earlier in this section with constant value of discharge coefficient (in this case Cd=1). The input to engine is angle of opening of throttle valve plate. The opening angle is measured with a plane perpendicular to the axis of pipe or air flow direction. The input angle is varied with accelerator pedal for several different values. The same input is fed to the derived model as input to evaluate its behavior and compare with manifold pressure measurements. It is clear that the derived model behaves very differently than the real engine operation.

Moreover, the manifold pressure value given by the model is very high with Cd=1; almost double the measurements throughout the experiment, except for a few points. At these points the model trajectory evolves in nearly opposite direction. It should be noted that for this simulation, the measured angular velocity of engine was used in model equation, and the rotational dynamics equation was not simulated. The similar results for second simulation experiment are shown in figure 3.2. Here, the value of Cd=0.5. As we can see that the lower value of Cd has brought the model manifold pressure trajectory significant low in the plot, and it almost proceeds closer to the engine measured inlet manifold pressure. But the evolution of both trajectories is not identical, which would have been; in case of a correct value of Cd. Both simulation experiments assert that the value of Cd must not be a constant, merely scaling the trajectory. But it must be a time varying parameter to correctly match the derived model to the engine measurements. The value of volumetric efficiency was taken to be 0.8 for model simulations. If engine measurements of volumetric efficiency are used in simulations, discharge coefficient would take different values.

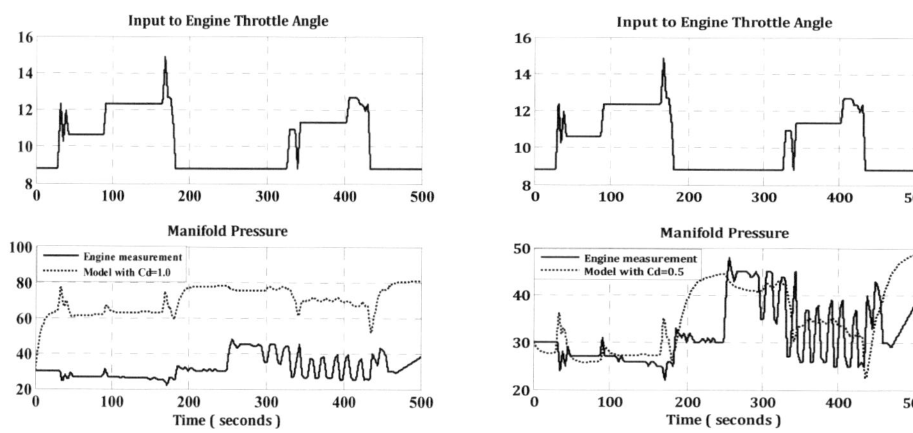

Figure 2. Throttle angle; above and manifold pressure; below with Cd=1.0 on left and Cd=0.5 on right. It is evident that the model trajectory is different than the engine measurements.

4. Hybrid model

Although the representation of SI engine as a hybrid model is already present in literature, the main difference of the approach presented in this thesis is the manner in which the continuous states of model are being represented. The hybrid model presented by Deligiannis V. F et al (2006, pp. 2991-2996) assumed the model of four engine processes [17] i.e. suction, compression, power and exhaust as four continuous sub-systems. Similar continuous systems are also considered in DEM that can also be considered as a hybrid model. In this model, each cylinders of engine is considered as independent subsystem that takes power generated due to the burning of air fuel mixture as input and movement of piston in engine cylinder is considered as the output. These sub-systems are represented as linear systems and complete SI engine is considered as a collection of subsystems. These subsystems are working coherently to produce the net engine output. The proposed hybrid model of SI engine can be regarded as a switched linear system. Although an SI engine is a highly nonlinear system, for certain control applications a simplified linear model is used. Li M. et al (2006, pp: 637-644) mentioned in [19] that modeling assumption of constant polar inertia for crankshaft, connecting rod and piston assemblies to develop a linear model is a reasonable assumption for a balanced engine having many cylinders. The modeling of sub-systems of proposed hybrid model would be performed under steady state conditions, when the velocity of system is fairly constant. Also the time in which the sub-system gives its output is sufficiently small. A linear approximation for modeling of sub-system can therefore be justified. Similar assumption of locally linear model is made by Isermann R et al (2001, pp: 566-582) in LOLIMOT structure [20]. The continuous cylinder dynamics is therefore represented by a second order transfer function with crankshaft speed as output and power acting on pistons of cylinder due to fuel ignition as input.

A continuous dynamic model of these sub-systems would be derived in this chapter. The timing of signals to fuel injectors, igniters, spark advance and other engine components is controlled by Electronic Control Unit (ECU) to ensure the generation of power in each cylinder in a deterministic and appropriate order. The formulation of hybrid modeling of sub-systems would be carried under the following set of assumptions:

Modeling Assumptions

1. Engine is operating under steady state condition at constant load.
2. Air fuel ratio is stoichiometric.
3. Air fuel mixture is burnt inside engine cylinder at the beginning of power stroke and energy is added instantaneously in cylinder resulting in increase in internal energy. This internal energy is changed to work at a constant rate and deliver energy to a storage element (flywheel).
4. At any time instant only one cylinder would receive input to become active and exerts force on piston and other cylinders being passive due to suction, compression and exhaust processes contribute to engine load torque.
5. All the four cylinders are identical and are mathematically represented by the same model

The switching logic can be represented as a function of state variables of systems.

4.1. Framework of hybrid model

The framework of Hybrid model for a maximally balanced SI engine with four cylinders is represented as a 5-tuple model $< \mu, X, \Gamma, \Sigma, \phi >$. The basic definition of model parameters is given below.

- $\mu = \{\mu_1, \mu_2, \mu_3, \mu_4\}$ where each element of set represents active subsystem of hybrid model.

- $X \in R^2$ represents the state variable of continuous subsystems, that would be defined when model is developed for subsystems, where the vector X consists of velocity and acceleration.
- $\Gamma = \{ M \}$ is a set that contains only a single element for a maximally balanced engine. M represents state space model of all subsystems and is assumed to be linear, minimum phase and stable. The model equation is derived in the next section. The model can be defined in state space as:

$$\dot{x}(t) = AX + BU \tag{27}$$

$$y(t) = CX + DU \tag{28}$$

Where

$$U \in R,\ A \in R^{2\times2},\ B \in R^{2\times1},\ C \in R^{1\times2},\ D \in R$$

- $\Sigma: \mu \rightarrow \mu$ represents the generator function that defines the next transition model. For an IC engine, the piston position has a one to one correspondence with crankshaft position during an ignition cycle. The generator function is therefore defined in terms of crankshaft position as:

$$\Sigma = \begin{cases} \mu_1 & 4n\pi \le \int \dot{\theta}_1 dt < (4n+1)\pi \\ \mu_2 & (4n+1)\pi \le \int \dot{\theta}_1 dt < (4n+2)\pi \\ \mu_3 & (4n+2)\pi \le \int \dot{\theta}_1 dt < (4n+3)\pi \\ \mu_4 & (4n+3)\pi \le \int \dot{\theta}_1 dt < (4n+4)\pi \end{cases} \tag{29}$$

  where n=0,1,2,… and $\int \dot{\theta}_1 dt$ represents instantaneous shaft position that identifies the output of generator function.

- $\phi: \Gamma \times \mu \times X \times u \rightarrow X$ defines initial condition for the next subsystem after the occurrence of a switching event, where u represents input to subsystem. Figure 4.1 shows the subsystems and switching sequence of proposed SI engine hybrid model.

4.2. Modeling of sub-system

A subsystem/cylinder is active when it contributes power to system i.e. during power stroke. When a sub-system is active its output is defined by the dynamic equations of system and its output during its inactive period is defined by its storage properties. The output of a sub-system provides initial condition to the next sub-system at the time of switching. All the subsystems are actuated sequentially during an ignition cycle. The cyclic actuation of subsystems is represented as a graph in Figure 4.1. The total output delivered by the system during complete ignition cycle would be the vector sum of outputs of all subsystems during that ignition cycle.

If T is the period of ignition cycle and u(t) is the input to system at time t within an ignition cycle and ui(t) is the input of ith subsystems; by assumption 4:

$$u_i(t) = u(t) \qquad when \ \frac{(i-1)T}{4} < t < \frac{iT}{4} \ , i = 1,2,3,4 \tag{30}$$

$$u_i(t) = 0 \qquad otherwise \tag{31}$$

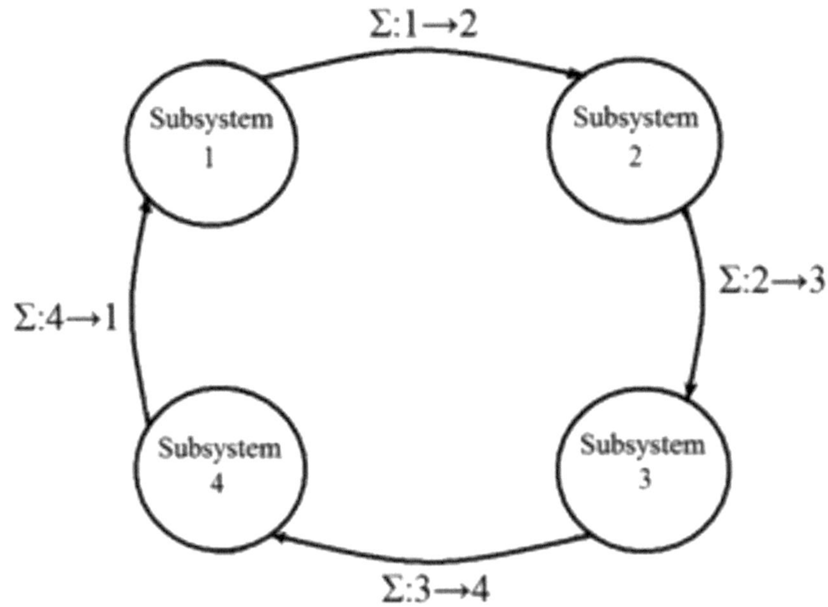

Figure 3. Switching of subsystems (*Adopted from Rizvi (2009, pp. 1-6*)

Franco et al (2008, pp: 338-361) used mass-elastic engine crank assembly model for real time brake torque estimation [24]. In this representation of SI engine each cylinder is represented by a second order mass spring damper as shown in Figure 4.2. Consider δQ amount of energy added in system by burning air fuel mixture. The instantaneous burning of fuel increases the internal energy δU in cylinder chamber.

$$\delta U = \delta Q \tag{32}$$

At ignition time, energy is added instantaneously in engine. This will increase internal energy of system. A part of this internal energy is used to do work and rest of the energy is drained in coolant and exhaust system. If internal energy change to work with constant efficiency ηt then work δW is given by the energy balance equation as:

$$\delta W = -\eta_t \, \delta U \tag{33}$$

Using equationg (27) we get

$$\delta W = -\eta_t \, \delta Q \tag{34}$$

If p is pressure due to burnt gases then work done during expansion stroke is given by:

$$W = \int_{V1}^{V2} p dV \tag{35}$$

where V1 and V2 are initial and final volume of cylinder during expansion. For adiabatic expansion:

$$pV^{\gamma} = k_1 \tag{36}$$

where k1 and γ are constant. Hence equation (30) becomes

$$W = \int_{V1}^{V2} k_1 V^{-\gamma} dV \tag{37}$$

$$W = k_1 \frac{V_2^{-\gamma+1} - V_1^{-\gamma+1}}{-\gamma+1} \tag{38}$$

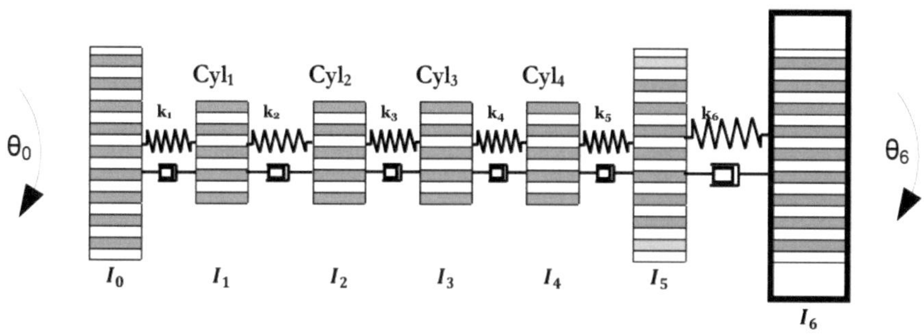

Figure 4. Spark ignition engine representation (Adopted from Franco et al (2008, pp. 338-361))

Consider that the closed end of the piston to be origin and x is a continuous variable representing the instantaneous piston position with respect to the origin. The piston always moves between two extreme positions xt and xb where xt represent piston position at Top Dead Center (TDC) and xb represent piston position at Bottom Dead Center (BDC). If the surface area of piston is A, and it moves a small distance δx from its initial position x, where δx is constant and can be chosen arbitrarily small, then using equation (33), work done can be expressed as:

$$\delta W = k_1 \frac{[A(x+\delta x)]^{-\gamma+1} - [Ax]^{-\gamma+1}}{-\gamma+1} \tag{39}$$

$$\delta W = k_1 \frac{A^{-\gamma+1}}{-\gamma+1} [(x+\delta x)^{-\gamma+1} - x^{-\gamma+1}] \tag{40}$$

$$\delta W = \frac{k_1 A^{-\gamma+1}}{-\gamma+1} \left[x^{-\gamma+1} \left(1 + \frac{\delta x}{x}\right)^{-\gamma+1} - x^{-\gamma+1} \right] \tag{41}$$

$$\delta W = \frac{k_1 A^{-\gamma+1} x^{-\gamma+1}}{-\gamma+1} \left[\left(1 + \frac{\delta x}{x}\right)^{-\gamma+1} - 1 \right] \tag{42}$$

Expanding using binomial series and neglecting higher powers of δx and simplifying:

$$\delta W = k_1 A^{-\gamma+1} x^{-\gamma} \delta x \tag{43}$$

Therefore from equation (29)

$$\delta Q - -\frac{k_1 A^{-\gamma+1} x^{-\gamma} \delta x}{\eta_t} \tag{44}$$

In deriving the model for sub-systems, each cylinder of SI engine is treated as a second order system as used in [24] and shown in Figure 4.2. Consider if F is the applied force by the burnt gases, m is the mass of engine moving assembly (piston, connecting rod, crankshaft and flywheel), coefficient of friction is k2 and coefficient of elasticity is k3, then net force acting on piston is given by:

$$m\frac{d^2x}{dt^2} = F - k_2\frac{dx}{dt} - k_3x \tag{45}$$

$$m\frac{d^2x}{dt^2} + k_2\frac{dx}{dt} + k_3x = F \tag{46}$$

Net work done by the expanding gases against the load, friction and elastic restoring forces when piston moves by a small distance δx would be given as:

$$\left[m\frac{d^2x}{dt^2} + k_2\frac{dx}{dt} + k_3x\right]\delta x = \delta W \tag{47}$$

Using equation (38) above equation becomes

$$\left[m\frac{d^2x}{dt^2} + k_2\frac{dx}{dt} + k_3x\right]\delta x = k_1A^{-\gamma+1}x^{-\gamma}\delta x \tag{48}$$

The displacement δx can be chosen constant and arbitrarily small. As the piston moves, the volume inside the combustion chamber increases resulting in the reduction of instantaneous pressure on piston. Instantaneous power is therefore a function of piston position. Instantaneous power delivered by the engine would be calculated by differentiation as:

$$\left[m\frac{d^3x}{dt^3} + k_2\frac{d^2x}{dt^2} + k_3\frac{dx}{dt}\right]\delta x = -k_1\gamma A^{-\gamma+1}x^{-\gamma-1}\frac{dx}{dt}\delta x \tag{49}$$

$$m\frac{d^3x}{dt^3} + k_2\frac{d^2x}{dt^2} + k_3\frac{dx}{dt} = -k_1\gamma A^{-\gamma+1}x^{-\gamma-1}\frac{dx}{dt} \tag{50}$$

Writing differential Eq 45 in terms of velocity v as:

$$m\frac{d^2v}{dt^2} + k_2\frac{dv}{dt} + k_3v = -\gamma\eta_t\frac{k_1A^{-\gamma+1}x^{-\gamma}\delta x}{\eta_t}\frac{v}{x\delta x} \tag{51}$$

$$m\frac{d^2v}{dt^2} + k_2\frac{dv}{dt} + k_3v = \gamma\eta_t\delta Q\frac{v}{x\delta x} \tag{52}$$

$$m\frac{d^2v}{dt^2} + k_2\frac{dv}{dt} + k_3v = \frac{\gamma\eta_t v}{x}\frac{\delta Q}{\delta t}\frac{\delta t}{\delta x} \tag{53}$$

$$m\frac{d^2v}{dt^2} + k_2\frac{dv}{dt} + k_3v = \frac{\gamma\eta_t v}{x}P(x)\frac{1}{v} \tag{54}$$

Assuming that crankshaft speed is proportional to the speed of piston inside the cylinder, equation (49) represents a model of crankshaft speed when energy is added in one of the cylinder of SI engine by the ignition of fuel. The model is however nonlinear on account of presence of x in the denominator on the right side of differential equation.

4.2.1. Model linearization

SI engine is a highly nonlinear system. In hybrid modeling, the time of activation of subsystems is very small. Also under under steady state conditions, the velocity of engine is fairly constant hence a linear approximation of engine subsystems can be justified. The model derived in earlier section is now linearized to form a switched linear model. The validity of linear model is only at the operating point. As x can never be zero, so the function is smooth and can be linearized at TDC. If the igniting fuel adds the power P(x) to a cylinder when piston is at position x, the dynamics of system at TDC would be described as:

$$m\frac{d^2v}{dt^2} + k_2\frac{dv}{dt} + k_3v = \frac{\gamma\eta_t}{x}P(x) \tag{55}$$

Linearizing the system at TDC ($x = x_t$) under steady state condition, and assuming that whole power is added in the cylinder instantaneously when the cylinder is at TDC, equation 50 becomes:

$$m\frac{d^2v}{dt^2} + k_2\frac{dv}{dt} + k_3v = \frac{\gamma\eta_t}{x_t}P(x) \tag{56}$$

In simulations, P(x) can be taken as a narrow pulse or a triangular wave, assuming that when system receives input, it deliver power at constant high rate for a short interval of time and thereafter the delivered power would be negligible. Since shaft speed is also constant at the start of each ignition cycle, therefore right hand side of equation 51 becomes constant and expression becomes a linear differential equation.

4.2.2. Model parameter estimation

The movement of piston exhibit a periodic behavior with same fundamental frequency as that of rotational speed of engine shaft. This provides a heuristic guideline to choose the value of k3 (in Eq 5.16) as a function of crankshaft angular speed. The empirical choice is validated using simulation and experimental results reported later.

$$k_3 = \omega^2 = (2\pi N)^2 \tag{57}$$

where N is engine speed in revolution per second.

During experimental verification load is also applied by friction. Most frictional models described in literature are based on empirical relations as a polynomial in engine speed. A simplified frictional model is chosen with term containing only square of engine speed. The constant term representing the load acting on engine is also considered as a parameter whose value is defined as a polynomial in crankshaft speed as:

$$k_2 = b\,\omega^2 + c \tag{58}$$

On the basis of simulation and experimental results it is established that the optimal selection of value of b varies between 0.02 and 0.5.

Parameter	Value	Description
m	20 Kg	Mass of Engine moving assembly
b	0.2	Friction Coefficient
k_3	10000	Elasticity Coefficient
γ	1.4	C_p / C_v
P	3 hp	Power generated in cylinder
η	0.3	Efficiency
ω	100 rad/s	Engine operating speed

Table 1. Parameter values used in simulation

4.2.3. Model properties and applications

The proposed hybrid model was used to study the properties of crankshaft speed of SI engine [Rizvi etal]. The simplicity of model also enabled to study some stochastic properties of engine variable also. An analysis of hybrid model indicates following results which are useful in statistical analysis of system.

1. Four peaks would be observed in one ignition cycle of a four cylinder SI engine.
2. Amplitude of four observed peaks represents four independent events.
3. Crankshaft speed is proportional to input power. (Due to linear model of subsystems)
4. Crankshaft speed is proportional to amount of intake air.

The model was used:

1. To develop state observer for estimation of angular acceleration
2. To detect and isolate the misfire fault in SI engine [22].

4.3. Model input estimation

The input to the model is the power generated inside the cylinder as a result of ignition. It is assumed that power operating on piston is coming from two sources i.e. by the ignition of fuel and by the power supplied by the engine rotating assembly due to inertia. In case of misfire, the power due to inertia of rotating assembly will maintain the movement of piston but the Power due to ignition of fuel is absent. Power can be defined as the product of force acting on piston of a cylinder and piston velocity. If F is the force acting on engine piston and v is the piston velocity, then power P acting on piston can be defined as:

$$P = Fv \tag{59}$$

$$P = p.A.v \tag{60}$$

Where p is the pressure inside the cylinder, A is the surface area of piston which is known. The only unknown variable is the cylinder pressure that can be estimated using observer or an estimator. One such technique of cylinder pressure estimation was proposed by Yaojung S. and Moskwa J. (1995, pp: 70-78) in [28]. However for simulation purpose typical values can be used. Under idle conditions the typical value of peak pressure inside the cylinders is 25 bars. If engine is running at 15 revolutions per second i.e. idle speed, and cylinder stroke is 75mm, then average speed of piston can be easily estimated. This pulse would be provided once in each ignition cycle i.e. in 720°. The time to traverse the complete stroke is 1/30 seconds or nearly 0.03 seconds. The average power provided by the fuel can now be estimated as:

$$\text{Power} = \left(\frac{2500000}{12} \times \text{pi} \times .075 \times .075/4\right) \times \left(\frac{.075}{.03}\right) \tag{61}$$

$$\text{Power} = 2301 \text{ Watt} = 3.1 \text{ hp}$$

A pulse with average value of power equivalent to 3.1 hp would then be used in simulations.

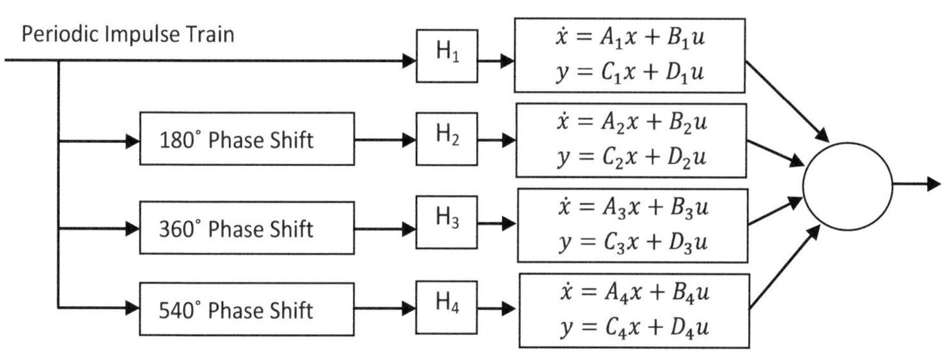

Figure 5. Switched linear system used for simulation purpose

4.4. Model simulation and experimental verification

The block diagram of switched linear system used for simulation purpose is shown in Figure 4.3. Input is provided as a periodic pulse train and three shifted versions of the same pulse train so that addition of all the four signals would also result in a periodic pulse train. H is a multiplier and represents health of a cylinder. H=1 represents a healthy cylinder that contribute to system output. H=0 represents faulty cylinder that does not contribute to system output.

4.4.1. Simulation results

For simulation computer program was written to implement the block diagram shown in Figure 4.3 in Matlab. This gain of all elements was given a value equal to 1 for no-misfire simulation. To simulate the misfire situation, the gain of the corresponding sub-system was set to zero so that its output did not participate in the net system output. The nominal values of model parameters/ constants used in simulation are provided in Table 1. Under no misfire condition, the model was tuned to match its output with the experimental results. Using same parameter values, the misfire situation was simulated. The simulation results of hybrid model for both healthy and faulty conditions are shown in Figure 4.4.

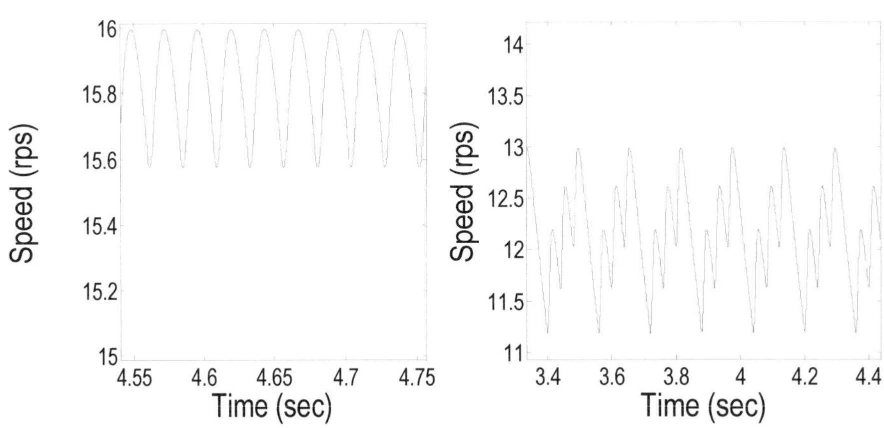

Figure 6. Simulation Results: The waveforms representing fully balanced engine operation (left) one cylinder misfiring (right)

The simulation results were then validated by conducting an experiment. In the crankshaft position was observed using crankshaft position sensor. The output of sensor is in the form of pulses. The data was logged using a data acquisition card from National Instrument Inc. on an analog channel with a constant data acquisition rate. Engine speed is estimated using crankshaft position data and experimental setup data:

$$\text{Number of Teeth in gear} = 13$$

$$\text{Angular spacing between normal Teeth} = 30°$$

$$\text{Angular spacing between double Teeth} = 15°$$

$$\text{Reference indication by Double teeth}$$

$$\text{Data Acquisition Rate} = 50000 \text{ samples/second}$$

The reference was first searched by finding the double teeth. The number of samples polled in the time interval of passing of two consecutive gear teeth in front of magnetic sensor was observed. The number of samples polled was converted to time as $\text{Time} = \frac{\text{Number of samples polled}}{\text{Data Acquisition Rate}}$.

Using angular displacement between two consecutive teeth and time to traverse that angular displacement, crankshaft speed was estimated. Crankshaft speed was finally plotted as a function of time. The experiment for the measurement of speed was conducted both under no-misfire condition and misfire condition. During experiment some load was kept on engine by application of brake. The value of applied load was however unknown but an effort was made to keep load similar in both experiments by retaining the brake paddle at the same position during both experiments. The experimental results are shown in Figure 7.

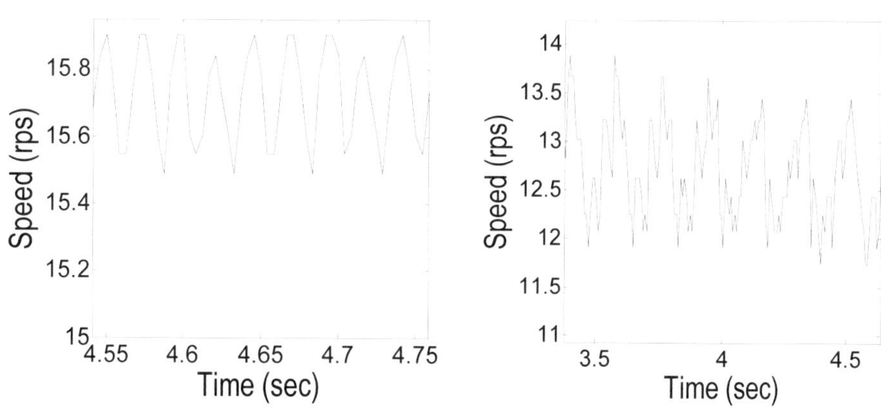

Figure 7. Experimental Results: The waveforms representing fully balanced engine operation (left) one cylinder misfiring (right)

Author details

Mudassar Abbas Rizvi and Aamer Iqbal Bhatti
Mohammed Ali Jinnah University (MAJU), Islamabad, Pakistan

Qarab Raza
Center for Advanced Studies in Engineering (CASE), Islamabad, Pakistan

Sajjad Zaidi
Pakistan Navy Engineering College, Karachi, Pakistan

Mansoor Khan
Jiao tong University, Shanghai, China

5. References

[1] L. Guzzella and C.H. Onder, "*Introduction to Modeling and Control of Internal Combustion Engine Systems*" Springer-Verlag Berlin Heidelberg 2004.
[2] John J. Moskwa "Automotive Engine modeling for Real time Control", PhD Thesis, Massachusetts Institute of Technology May, 1988.
[3] Elbert Hendricks, Spencer C. Sorenson, "Mean Value Modeling of Spark Ignition Engines," SAE Technical Paper no.900616, 1990.
[4] Elbert Hendricks, Alain Chevalier, Michael Jensen and Spencer C. Sorensen "Modeling of the Intake Manifold Filling Dynamics", 1996. SAE Technical Paper No. 960037.
[5] Elbert Hendricks, "Engine Modeling for Control Applications: A Critical Survey," Meccanica 32: 387-396, 1997.
[6] Crossly, P., R., Cook, J., A.,"A Nonlinear Engine Model for Drive train System Development," IEEE Int. Conf. 'Control 91',Conference Publication No.332, Vol.2.,Edinburgh,U.K.,March 1991.
[7] Dobner, Donald J., "A Mathematical Engine Model for Development of Dynamic Engine Control," SAE technical paper No. 800054, 1980.
[8] Gordon P. Blair, "*Design and Simulation of Four Stroke Engines*", SAE International, 1999. ISBN 0-7680-04440-3.
[9] Weeks, R, W, Moskwa, J, J, "Automotive Engine Modeling for Real-Time Control Using Matlab/Simulink," SAE Technical Paper no.950417, 1995.
[10] Robert Todd Chang, "A Modeling Study of the Influence of Spark-Ignition Engine Design Parameters on Engine Thermal Efficiency and Performance", Master's Thesis, Department of Mechanical Engineering, Massachusetts Institute of Technology, 1988.
[11] Q. R. Butt, A. I. Bhatti, M. Iqbal, M. A. Rizvi, R. Mufti, I. H. Kazmi, "Estimation of Gasoline Engine Parameters Part I: Discharge Coefficient of Throttle Body", IBCAST 2009, Islamabad.
[12] Q. R. Butt, A. I. Bhatti, "Estimation of Gasoline Engine Parameters using Higher Order Sliding Mode", IEEE Transaction on Industrial Electronics *Vol* 55 Issue 11 Nov 2008 Page 3891-3898.

[13] R. Scattolini, A. Miotti, G. Lorini, P. Bolzern, P., Colaneri, N. Schiavoni, "Modeling, simulation and control of an automotive gasoline engine" Proceedings of 2006 IEEE, International conference on control applications, Munich, Germany, October 4-6, 2006.
[14] Toshihiro Aono and Takehiko Kowatari, "Throttle-Control Algorithm for Improving Engine Response Based on Air-Intake Model and Throttle-Response Model", IEEE Transactions in Industrial Electronics, vol. 53, No.3, June 2006.
[15] Christian Bohn, Thomas Bohme, Aik Staate and Petra Manemann, "A Nonlinear Model for Design and Simulation of Automotive Idle Speed Control Strategies", Proceeding of the American Control Conference 2006, Minneapolis, Minnesota, USA, June 14-16 2006.
[16] Per Andersson "Air Charge Estimation in Turbocharged Spark Ignition Engines", Doctoral Thesis, Department of Electrical Engineering, Linkoping University, Linkoping, Sweden 2005.
[17] Deligiannis V, Manesis,S : "Modeling Internal Combustion Engines Using a Hyper-Class of Hybrid Automata: A Case Study", Conference on Computer Aided Control Systems Design, Munich, Germany, Proceedings of the 2006 IEEE, pp. 2991-2996
[18] Sengupta, S., Mukhopadhyay, S., Deb, A., Pattada, K. et al., "Hybrid Automata Modeling of SI Gasoline Engines towards state estimation for fault diagnosis," SAE Technical Paper 2011-01-2434, 2011, doi:10.4271/2011-01-2434
[19] Wong P. K, Tam L. M., Li K, Vong C. M, "Engine Idle Speed System Modeling and Control Optimization using Artificial intelligence", Proceedings of the Institution of Mechanical Engineers. Part D, Journal of Automobile Engineering, Volume 224, 2009, pp: 55-72.
[20] Isermann R, Müller N., "Modeling and Adaptive Control of Combustion Engines With Fast Neural Networks", European Symposium on Intelligent Technologies, Hybrid Systems and their Implementation on Smart Adaptive Systems, Tenerife, Spain, 2001.
[21] Rizvi M. A., A. I. Bhatti, "Hybrid Model for Early Detection of Misfire Fault in SI Engines", IEEE 13th International Multitopic Conference, 538315, pp 1-6, Nov. 2009.
[22] Rizvi M. A, Bhatti A. I., Butt Q. R, "Hybrid Model of Gasoline Engine for Misfire Detection", IEEE Transactions on Industrial Electronics" Accepted for publication in 2010.
[23] Karlsson J, Fredriksson J. "Cylinder-by-Cylinder Engine Models Vs Mean Value Engine Models for use in Powertrain Control Applications", Society of Automotive Engineers, SAE 1998. 99P-174, pp: 1-8
[24] Franco J, Franchek M. A, Grigoriadis K, "Real-time brake torque estimation for internal combustion engines", Mechanical Systems and Signal processing, 22 (2008), pp: 338-361
[25] Matco, D., Zupancic, B., Karba, R., "Simulation and Modeling of Continuous System, A Case Study Approach," Prentice Hall, 1992.
[26] Heywood; John B., "Internal Combustion Engine Fundamentals," McGraw-Hill Inc., 1988.

[27] Per Andersson, Lars Eriksson and Lars Nielsen "Modeling and Architecture Examples of Model Based Engine Control" Vehicular Systems, ISY, Linköping University, Sweden.

[28] Yaojung Shiao, John J. Moskwa, "Model-Based Cylinder-By-Cylinder Air-Fuel Ratio Control for SI Engines using Sliding Observers" Proceedings of the 1996 IEEE International Conference on Control Application Dearborn, MI. September 15-18, 1996.

Chapter 8

The Study of Inflow Improvement in Spark Engines by Using New Concepts of Air Filters

Sorin Rațiu and Corneliu Birtok-Băneasă

Additional information is available at the end of the chapter

http://dx.doi.org/10.5772/48137

1. Introduction

The piston internal combustion engine is a thermal engine that converts the chemical energy of the engine fluid fuel into mechanical energy. Engine fluid developments are achieved by means of a piston. The alternative movement of the piston inside a cylinder becomes rotating movement due to the crank gear.

For an internal combustion engine, the gas changing process encloses the intake and exhaust, which condition each other. The intake process is the process during which fresh fluid (air) enters the engine cylinders. The intake (or filling) determines the amount of fresh fluid retained in the cylinder after closing the last filling body and thus the mechanical energy developed during relaxation. The exhaust determines the purification degree of the cylinder with a view to a subsequent fill. In other words, the bigger the amount of fresh fluid (respectively air) retained into the engine cylinders, the higher the engine performance. The large amount of air in the engine cylinders means high pressure and low temperature on the inlet. This is the origin of the idea for this study which seeks ways to maximize, as much as possible, the amount of air introduced into the engine cylinders (by increasing pressure and decreasing temperature into cylinders), during an operating cycle, the costs for this goal being minimal.

Cylinder filling can be normal or forced (supercharging).

Normal filling, or normal inlet, typical of only 4-stroke engines, is achieved due to the piston's movement in the cylinder, in the sense of volume increase. Volume growth is recorded in the intake stroke, the fresh fluid with atmospheric pressure on the inlet.

Forced filling is achieved when the inlet pressure is greater than atmospheric pressure, indispensable in the 2-stroke engine without gas exchange bound drives. Forced filling can be achieved by supercharging when special equipment prepares the fresh fluid to enter the engine inlet at a pressure greater than the atmospheric one.

The cylinder filling process is strongly influenced by gas-dynamic losses on the engine intake route. There are two kinds of losses:

- *Thermal losses* due to heating the fluid through the inlet route walls, thus the final temperature being T + ΔT, the temperature increase resulting in diminishing density and hence the filling penalty.
- *Pressure losses* caused by the existence of hydraulic resistances on the intake route and fluid friction with the pipe walls. These can be quantified according to the well-known formula [1]:

$$\Delta p = \xi \cdot \rho \cdot \frac{w^2}{2} \tag{1}$$

where ξ is the pressure loss coefficient, w is the flow speed of the fresh fluid and ρ is its density.

Due to these losses, the amount of fresh charge retained in the engine cylinders, while providing information on the filling conditions, cannot serve as a comparison standard for different engines, but only for the same engine (the size of the losses mentioned above differs from one engine to another). This is why we introduce the notion of filling degree, or filling coefficient, or filling efficiency as a criterion for assessing filling perfection [1]:

$$\eta_v = \frac{C}{C_0} \tag{2}$$

where C is the amount of fresh fluid actually retained in the cylinder, and C_0 is the amount of fresh fluid that could be retained in the cylinder, under the state conditions of the engine inlet, i.e. without taking into account the losses mentioned above.

Proper filtering of the air that enters the internal combustion engine cylinder is essential to extend its operation. Preventing the intake of various impurities along with atmosphere air significantly reduces the wear and tear of engine parts in relative movement.

Unfortunately, besides the function of filtering air drawn from the atmosphere, the air filter – as a distinct part in engine composition - is a significant gas-dynamic resistance interposed on the suction route. If it is not cleaned regularly and the vehicle is driven frequently in dusty areas, the suction pressure p_a is reduced consistently and the filling efficiency η_v suffers penalties [1].

The air filters for filtrating sucked air, needed for the running of internal combustion engines, are made in several variants that differ according to the filtration principle:

- filters with filter element;
- filters by inertia;
- combined filters.

These variants have the following disadvantages:

- the existence of the filter element inside the case marks increased gas-dynamic resistance on the absorbed air (resulting in insufficient absorption phenomenon);

- the storage of dirt inside the filter interferes with the self-cleaning property of the filter element;
- the inability to view the filter element without prior removal of the filter to verify its impurity loading degree;
- the inability of the air filter to considerably increase the speed of the absorbed air;
- the inability of the air filter to pre-cool the air drawn into it;
- the inability of the air filter to create a slight boost while driving.

Air filtration is characterized by the following parameters:

1. average retention efficiency - measure of the filter's ability to capture dust and other impurities from the filtered air, expressed in [%];
2. filter class - the filter material's ability to retain particles of minimum specified size;

Composition of dust particles in the air, for example see [2]:

Airborne dust is extremely variable both as component material and particle size, depending on geographical area and climate. In general, airborne dust contains SiO_2, CaO_2, MgO, Fe_2O_3, Al_2O_3, alkaline material, organic material, soot, debris and smoke. Most dangerous for the engine are quartz particles (sand), quartz being one of the strongest abrasives. Organic substances are harmless to the engine.

The density of dust in the air is between 0.0002 g/m^3 in winter and 3-4 g/m^3 in summer, on earth roads from 0.5 to 1 m height above the ground. On paved roads, in summer, dust content in the air is 0.001 to 0.002 g/m^3, i.e. 20 to 30 times higher than in winter; the density variation between the minimum and maximum, i.e. from 0.0002 to 4 g/m^3, is 1 to 20 000. If during daytime you can see only up to 25 to 30 m, there is about 1g dust/m^3 of air.

The most dangerous particles are quartz grains exceeding 0,010 mm, quartz being a powerful abrasive which causes wear and tear 2-3 times more than smaller particles. There is no air filter to stop all the dust in the air, meaning one with 100% filtration efficiency.

2. Brief history of the evolution of air filters

At the beginning of the internal combustion engine evolution, the air filter's role was limited to the simple function of filtering the air entering the cylinders, air absorption taking place only from the engine compartment, regardless of season. Filter element was made of stainless steel sieves overlaid in 5 to 10 layers. Filter shape, respectively of the filter element, was mostly cylindrical (Figures 1-4) [2].

Figure 1. FORD T Model 1928

Figure 2. FORD A Model 1929

Figure 3. FORD V8 1932

Figure 4. FORD V8 1932

In certain racing engines with carburetors mounted on each cylinder, there is even no air filter, the engine sucking air directly from the engine compartment, without pre-filtering it (Figure 5).

Figure 5. MILLER V16 1931

The filter element becomes a consumable item, being replaced at road running intervals set by each manufacturer.

After the '40s, differential absorption of air depending on season starts to appear: in summer outside the engine compartment and in winter from the exhaust manifold area (Figures 6-11) [2].

The filter element known today as microporous or textile cardboard appears much later, in a variety of forms, the most common being the panel (Figure 12).

Figure 6. . FORD MUSTANG I 1964

Figure 7. FORD MUSTANG II 1974

Figure 8. FORD MUSTANG GT 1982

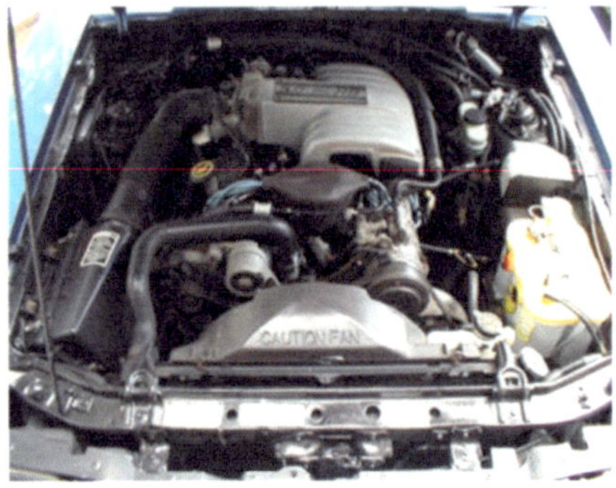

Figure 9. FORD MUSTANG 1994

Figure 10. FORD MUSTANG 2001

Figure 11. FORD MUSTANG 2005

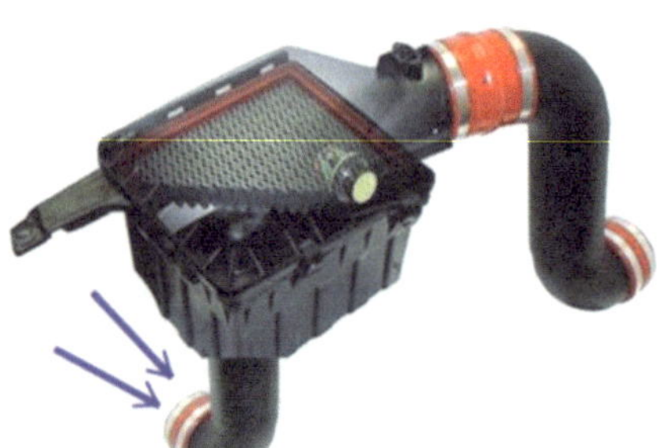

Figure 12. Panel type air filter

3. Super absorbing air filters (SAAF) – Own concepts [2,5]

From the outset it should be noted that the concepts proposed by the authors refer exclusively to air filter casing. The filter element is standard, purchased from well-known manufacturers.

The super absorbing or multifunction filters designed by the authors fulfil, besides the main task of filtering the air, the following functions:

- To capture the air;
- To increase the air flow rate of the absorbed air;
- To pre-cool the air;
- To change the airflow direction;
- To increase the filling coefficient or efficiency.

Here follows a classification of the super absorbing filters depending on assembly position, gauge dimensions and air intake function.

3.1. Super absorbing cylindrical air filter with internal diffuser (SAAFid)

The internal diffuser has the additional function of accelerating air speed out of the filter. The design geometry provides considerable increase in the coefficient of filling the cylinders with air.

The internal diffuser has variable sizes, depending on engine displacement. The larger the displacement is, the higher the diffuser sizes are, and vice versa.

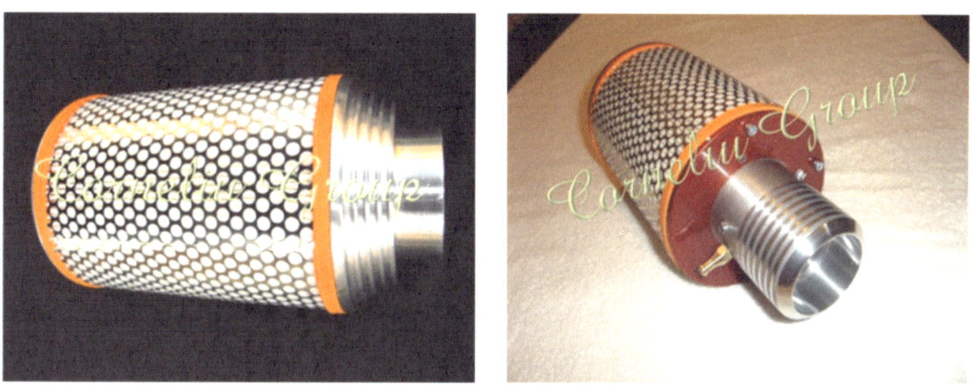

(a) (b)

Figure 13. Super absorbing filter with internal diffuser: a, b – physical models

3.2. Supliform super absorbing air filter (suSAAF)

It is mainly based on space-saving, reduced gauge dimensions (flexibility), being useful where the engine compartment is very tight (large capacity engines with supercharging installations). This filter allows, in its turn, function multiplication, namely, besides the main function of filtering the air, the filter increases the absorption and intake rate, as well as the speed of the absorbed air, also cooling it. This filter type can be a filter with one filter area and external collector, and a filter with dual filtration area.

The supliform air filter with one filter area consists of an axial external collector of cylindrical-concave shape (in radial section it is an arc), lined with filter element, concentration surfaces and connection cylinder (figure14).

The arc-shaped axial external sensor is radially closed at both ends with two concentration surfaces (crescent shaped) (Figures 14.c).

The filter element is recessed axially between the collector edges and radially between the concentration surfaces.

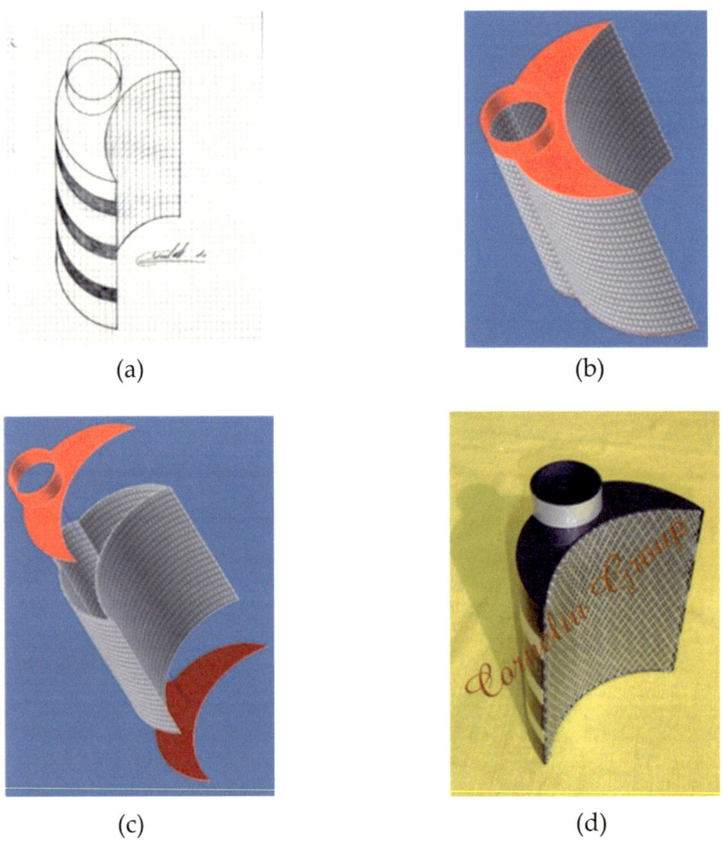

a – sketch, b – virtual model made in AutoDesk Inventor, c – exploded virtual model, d – physical model

Figure 14. Supliform super absorbing air filter with one filter area and external collector

On one of the concentration surfaces there is the fitting cylinder (Figure 14.d), which provides connection of the supliform filter to the engine inlet.

The collector, together with the concentration surfaces, captures radially and directs axially the air into the fitting cylinder (towards filter exit).

Due to its geometric shape, the collector ensures the minimum gas-dynamic resistance and creates a slight boost which increases proportionally with the speed of the vehicle, increasing considerably the amount of air absorbed and thus the filling coefficient of the engine.

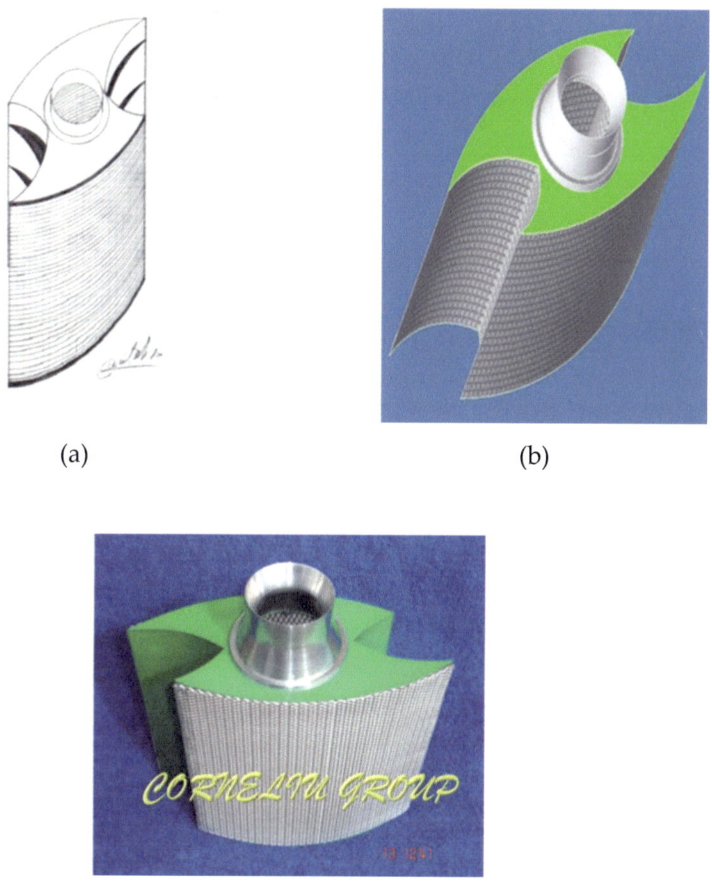

(a) (b)

(c)

a – sketch, b – virtual model made in AutoDesk Inventor, c – physical model

Figure 15. Supliform super absorbing air filter with dual filtration area

The filter element has a concave semi-cylindrical shape (Figure 14.d) and defines the area located between the collector edges and concentration surfaces. It is made of micron cardboard placed so that it forms the lateral area of the filter element (in radial section the micron cardboard has a VV shape). The cardboard provides a fine filter (microns) and is covered on the outside with a millimeter sieve, which allows coarse air filtering (millimeter). The micrometer cardboard and millimeter sieve are rigid (fixed) at both open ends with silicone semi-rings for better alignment and ideal tightness relative to the sensor edges and concentration surfaces.

The supliform air filter is mounted radially to the geometric axis of the vehicle (perpendicular on travel direction) for proper intake and absorption performance (Figure 16).

Figure 16. Mounting supliform air filter on engine

3.3. Super absorbing air filter with wide filtration range (SAAFwr)

This type of filter is made up of a front diffuser and a side surface with guiding cells, allowing the capture of both front and side air, thus increasing the surface of air penetration into the filter.

Figure 17. Super absorbing air filter with wide filtration range

4. Dynamic air transfer device (DATD) [2,5]

During operation of internal combustion engines fitted on motor vehicles in summer, there can be noted two shortcomings of the super absorbing filters, leading to their poor performance:

- insufficient air suction effect due to their installation in the engine compartment, area where the airflow is turbulent, the airflow around the filter not being laminar but turbulent, and the volumetric efficiency suffering penalties;
- increase in the absorbed air temperature due to their installation in the engine compartment, area which is subject to thermal radiation (of the cooling radiator and exhaust manifold). The overheating of the fresh fluid increases engine temperature, the appearance of detonation combustion and engine power reduction.

Figure 18 presents the most disadvantageous fitting solutions in terms of exposure to thermal radiation and eddy currents.

(a)

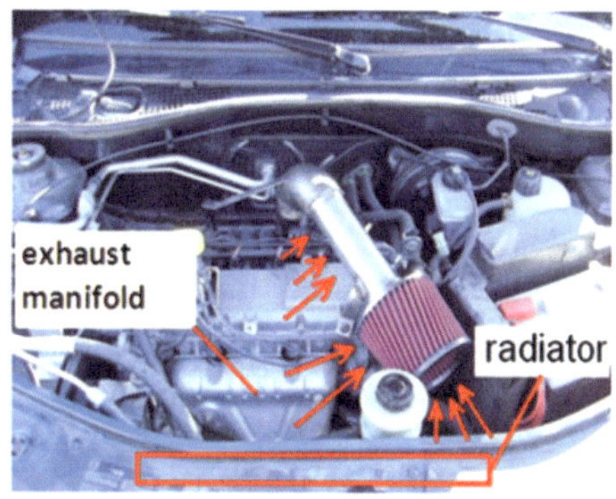

(b)

Figure 18. Assemblies that lead to direct exposure of the filters

It was tried to use a separator compartment for the air filter (Figure 19). But the separator compartment does not provide ideal insulation against the engine compartment, allowing thermal radiation to enter the air filter. Due to its design, turbulent air currents are created inside the separator compartment.

(a)

(b)

Figure 19. Installing filters in separator compartments

Moreover, the extension of the intake route outside the engine compartment is also practised. The example is given by mounting the air filter in the front bumper of a Honda

Civic R Type. At the same time, the incorporation of the air filter in aluminum or carbon cage is also practised.

By using the above solutions, the disadvantages mentioned are partially eliminated, but there is significant pressure loss due to extension of the intake route and existence of casting. Suction pressure is reduced consistently. On average, the intake route distances increase by over 500 mm, leading to increased drag created by additional friction arising from contact with the intake route wall.

Considering the above drawbacks, an efficient intake device was designed for internal combustion engines, called *dynamic air transfer device* (DATD) (Figure 20). It helps improve air circulation to the filter air through the engine compartment.

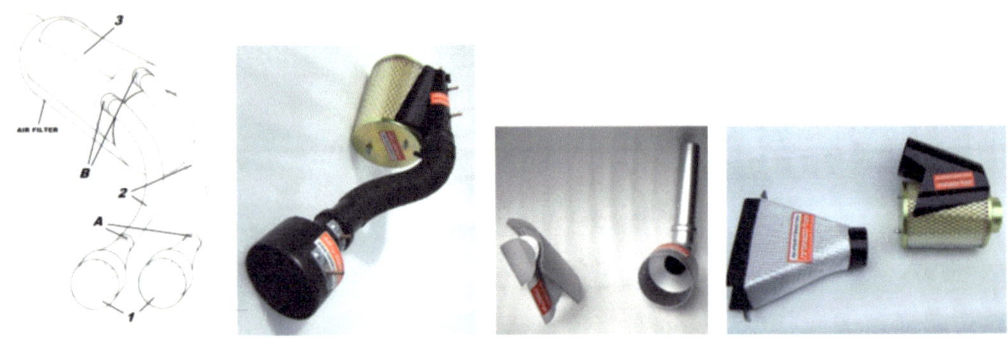

(a) (b) (c) (d)

a - design scheme: 1 - external collector diffuser, 2 - pipe connection, 3 - axial external collector, A, B - connecting surfaces; b - real overview; c - axial external collector and external collector diffuser - overview; d - different types of axial external collector and external collector diffuser.

Figure 20. Dynamic air transfer device (DATD)

The novelty consists in mounting external collector diffusers longitudinally with the vehicle axis, in the front area (in front of the radiator area or in the front bumper). They drive the air trapped outside the engine compartment, through the pipe connection, to the axial external collector, where the transfer to the air filter takes place.

The dynamic air transfer device consists of:

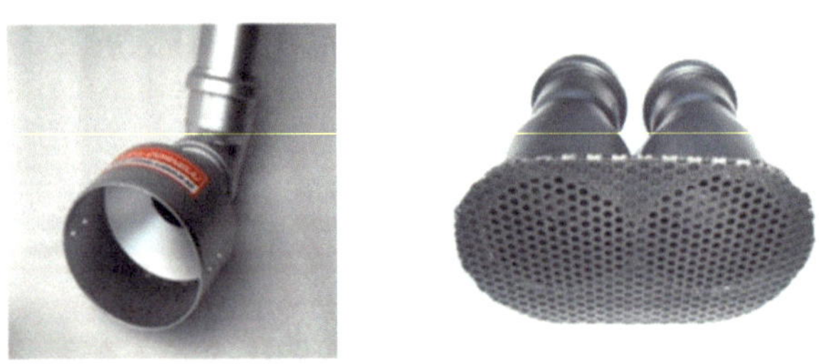

(a) (b)

Figure 21. External collector diffusers - physical models

1. External collector diffusers (one or more), whose role is to capture and accelerate air velocity (Figure 21).
2. Pipe connection, which connect the external collector diffusers and axial external collector (Figure 22).

Figure 22. Pipe connection

3. The axial external collector (mono or bi-route), Figure 23, is mounted on the super absorbing air filter oriented to the high heat radiation areas (exhaust manifold, radiator, engine). It takes at least 30% of the lateral filter area, being at a well-determined distance away from the filter (between 3 and 8 mm). Its role is to transfer the air flow in the filter, flow divided into two components: one that actually enters the filter, the actual flow being admitted into the engine cylinders and one that surrounds the lateral surface of the filter, leading to keeping a relatively low filter temperature.

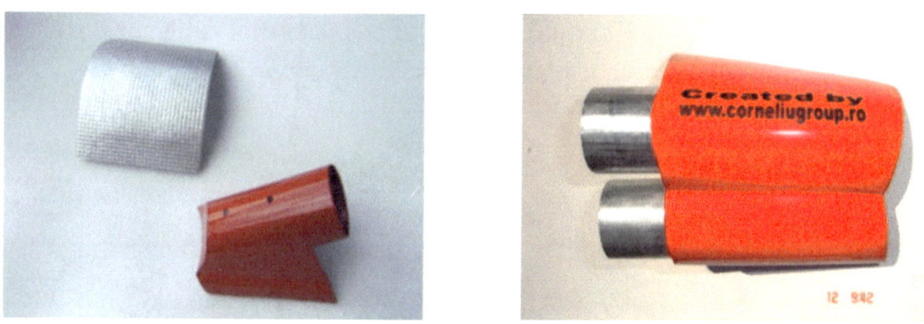

(a) (b)

a – mono-route, b – bi-route

Figure 23. Axial external collector

While driving the vehicle, the air is taken over by the external collector diffusers which enhance its speed, concentrate and convey it through the pipe connection to the axial external collector that transfers it to the super absorbing air filter.

The air stream transferred (brought) from outside the engine compartment has laminar focused flow. Speed increases (task performed by the external collector diffusers), and at the same time the air flow temperature decreases significantly. Because of its design geometry, the axial external sensor ensures good transfer and dispersion of the air on the side area of the super absorbing air filter. The amount of transfered air increases proportionally with the speed of the vehicle.

(a) (b)

Figure 24. a - DATD mounted on the engine; b - component parts: 1 - external collector diffusers, 2 - pipe connection, 3 - bi-route axial external collector

Advantages of DATD:

- the air transfer to the filter has laminar focused flow;
- the low air temperature provides improved filling efficiency;
- a slight boost is created increasing proportionally with the speed of the vehicle;
- the combustion process is improved;
- the tendency is toward dynamic inlet;
- it allows shortening the distance between the filter and the intake manifold.

Depending on engine capacity, one should use one or two external collector diffusers and an axial external collector in one or two transfer routes with varying sizes.

4.1. Experimented DATD

Figure 25. DATD mounted on Renault LAGUNA 1.6 16V

Differential pressure measurements were made both in the presence of the axial external collector of the DATD, and in its absence, in the suction area of the air filter, on different speed ranges. Data were taken while driving in real traffic at different speeds of the vehicle.

Figure 26. DATD mounted WV Golf 5, 1.4 16v

Figure 27. DATD, bi-route mounted on Honda Civic, 1.6

Figure 28. DATD, mounted on Renault Megane Coupe

For measurements in the transfer area of the axial external collector - air filter, the pressure intake port was oriented axially to the airflow direction.

For measurements without DATD, in the suction area of the air filter, the pressure intake port is directed axially to the air absorption direction.

One can notice a higher air capture and transfer effect obtained by DATD, in comparison with the simple absorption of the super absorbing air filter. This effect is accentuated by higher values of the relative air pressure in the capture area obtained by DATD, in comparison with the relative pressure values in the absorption area of the air filter in the absence of DATD. This is particularly beneficial, especially for non-supercharged engine intake (normal inlet), the amount of air admitted into the engine cylinders being directly proportional to the intake pressure.

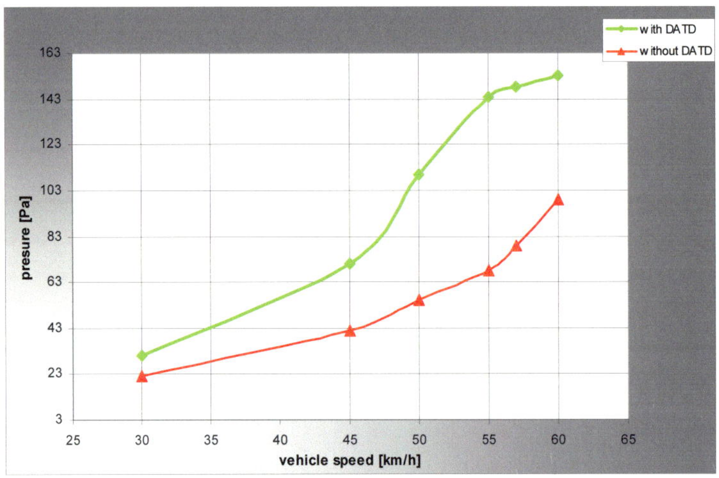

Figure 29. Comparative graph of the relative air pressure fields in the suction area of the air filter, in the presence and respectively absence of DATD

Here are some comparative measurements of the temperatures of the outer surfaces of the original classic air filter (OAF) and super absorbing air filters with DATD (SAAF+DATD) in real traffic conditions.

From the graphs above it can be seen that the temperatures of the outer surfaces of the super absorbing filters, in combination with the dynamic air transfer device (SAAF+DATD), are much lower than the ones corresponding to the outer surfaces of the original air filters (OAF). Consequently, the use of super absorbing air filters together with DATD leads to less acute air heating on the intake route and therefore improvement of the filling coefficient.

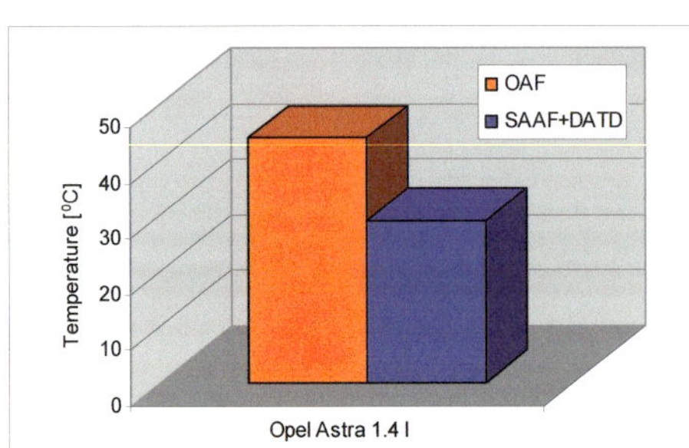

Figure 30. Temperatures of the outer surfaces of the OAF and SAAF+DATD for Opela Astra 1.4i

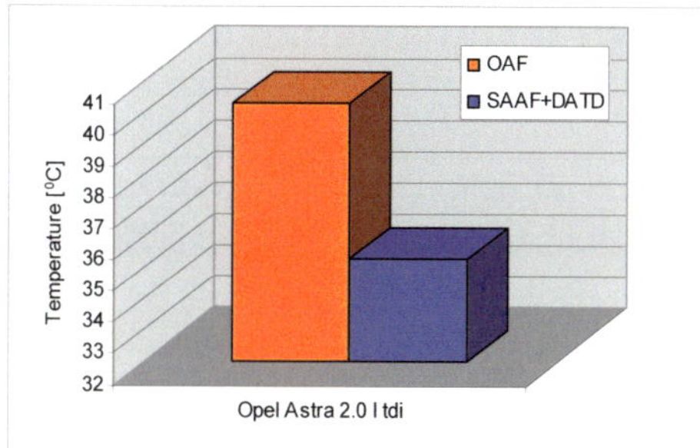

Figure 31. Temperatures of the outer surfaces of the OAF and SAAF+DATD for Opela Astra 2.0tdi

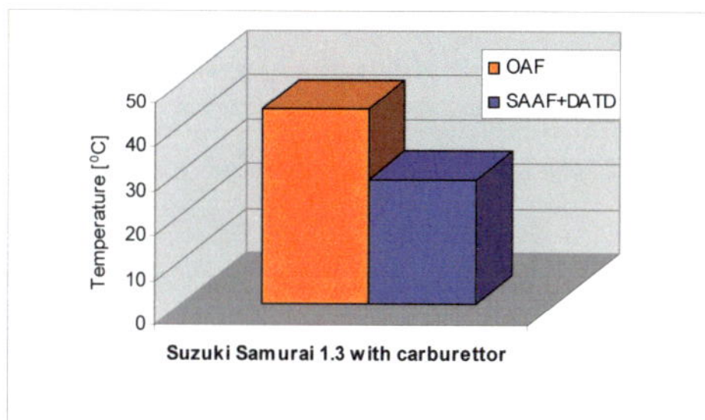

Figure 32. Temperatures of the outer surfaces of the OAF and SAAF+DATD for Suzuki Samurai

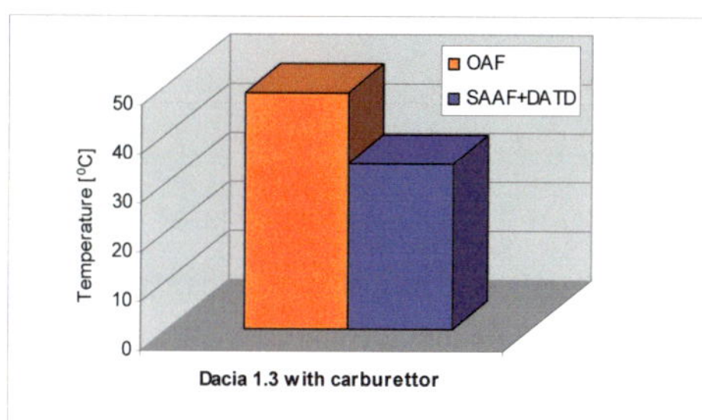

Figure 33. Temperatures of the outer surfaces of the OAF and SAAF+DATD for Dacia 1.3

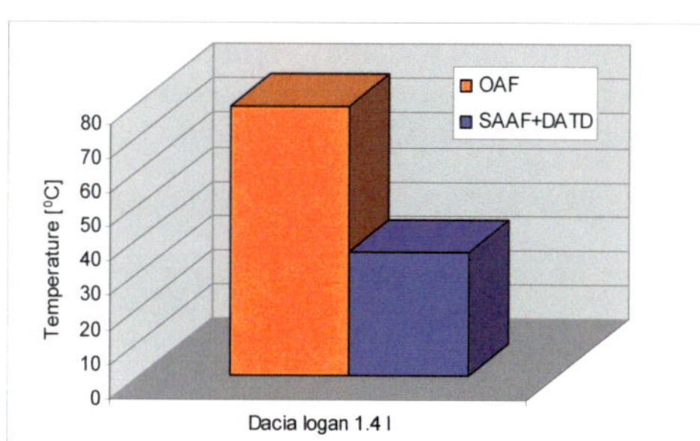

Figure 34. Temperatures of the outer surfaces of the OAF and SAAF+DATD for Dacia Logan 1.4i

5. Integrated deflector (ID) for attenuation of thermal radiation coming from the cooling radiator [2]

The thermal radiation and warm air from the engine cooling radiator extra heat the air filter and intake manifold. The absorbed air is also heated thus decreasing its density, the engine performance diminishing especially in hot weather. The air filter and intake manifold temperatures vary, in this case, between 60 and 85 °C, depending on the car speed.

Figure 35. Illustration of thermal radiation orientation towards the air filter

Figure 36. Filter assembly without the cooling radiator's integrated deflector

The cooling radiator's integrated deflector is designed to reduce these shortcomings, being mounted behind the radiator fan to direct the air flow beneath the inlet level (downwards). The deflector is thermally insulated (Figure 38), the filter and manifold temperatures falling within the range 25 ... 37 °C when the deflector is used.

Figure 37. Radiator fan

Figure 38. Integrated deflector - physical model

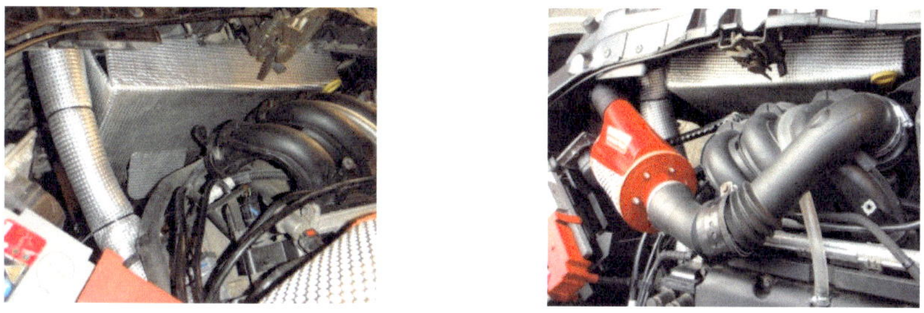

(a) (b)

Figure 39. a,b - Overview of radiator with mounted deflector

As already mentioned, the purpose of the integrated deflector is to direct downward the hot airflow passing through the engine cooling radiator (Fig. 40, b).

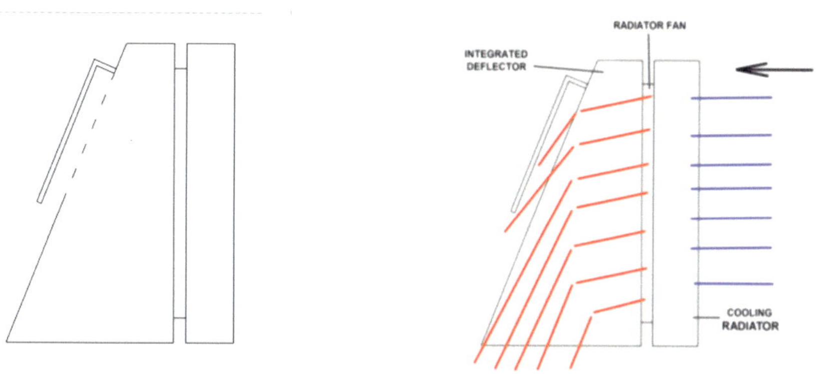

(a) (b)

Figure 40. Integrated deflector: a - sketch, b - operation principle

The technical problem solved consists in protecting the intake manifold and air filter from the heat radiation coming from the engine cooling radiator.

By use of the deflector integrated the following advantages are obtained:

- downward direction of the hot airflow coming from the cooling radiator (thermal radiation), outside the engine compartment,

- maintaining an optimum temperature of the intake manifold and air filter (to avoid overheating them).

The integrated deflector is provided with a deflector wall (Figure 41), which has a rectangular concentration area (2) fixed on the upper end. The concentration trapezoidal surfaces (3) and (4) are fixed on the lateral ends of the deflector wall (1), with the large trapeze end at the bottom. The deflector wall (1) has two or more directional windows (5).

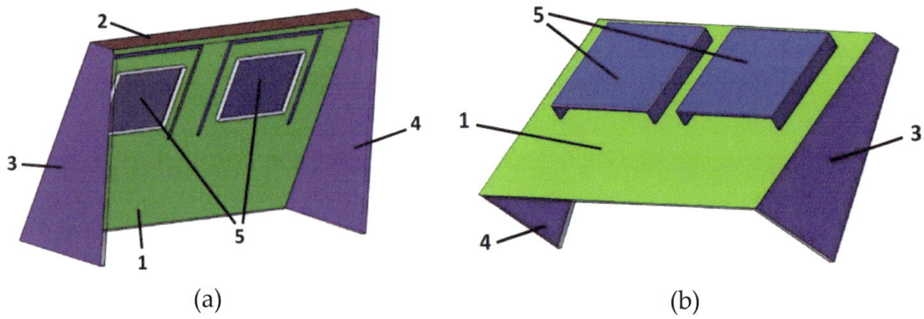

Figure 41. Integrated deflector - virtual model, made in Autodesk Inventor:
1 - deflector wall, 2 - rectangular concentration area;
3, 4 - trapezoidal concentration surfaces; 5 - directional windows

The bottom surface between bases (3), (4) and the bottom edge of the deflector wall (1) is open (free) to allow the evacuation of most hot airflow coming from the cooling radiator. The directional windows (5) allow additional exhaust of the hot airflow coming from the cooling radiator.

The deflector is not bad for engine cooling, the operating temperature of the coolant remaining within normal operating parameters.

Further experimental measurements are shown for comparative temperatures of intake air in the presence and absence of the deflector. Please note that in summer tests were made on

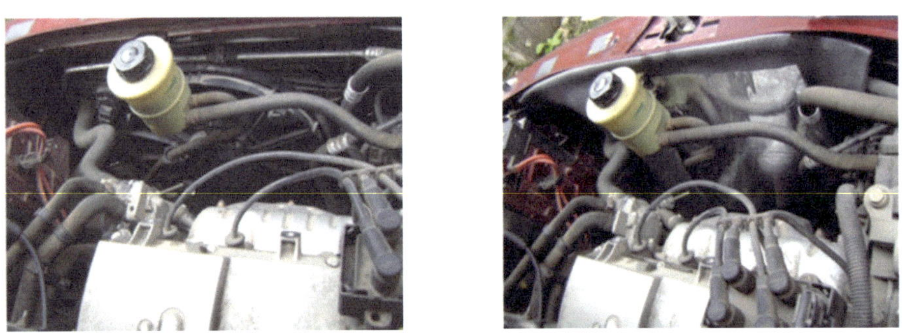

(a) (b)

Figure 42. Overview of engine radiator tested: a - without integrated deflector, b - with integrated deflector mounted

4 different cars, drawing the conclusion that the deflector has no adverse effect on engine cooling, the operating temperature of the coolant remaining within the parameters specified by the manufacturer.

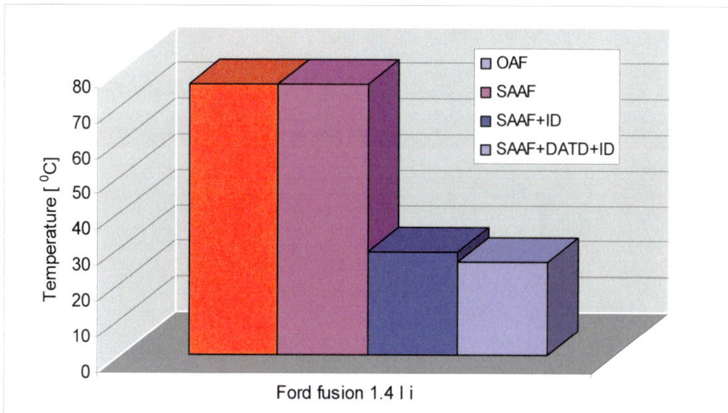

Figure 43. Intake air temperature values in the proximity of the air filter

As illustrated in Figure 45, the intake air temperature values for the original air filter (OAF) and the super absorbing air filter (SAAF) are similar in size and relatively high, leading to low density of the fresh load in the cylinders and thus to reducing the filling efficiency. Conversely, the temperature values recorded in the presence of the super absorbing air filter with integrated deflector (SAAF+ID) and to which the dynamic air transfer device is added (SAAF+DADT+ID) are much lower than the previous ones, which favors the improvement of the filling efficiency.

In conclusion, we can say that the dynamic air transfer device together with the integrated heat deflector, lead on the one hand to increasing the fresh fluid intake pressure, and on the other hand to lowering its temperature, both solutions contributing to increasing the filling efficiency η_v of the engine cylinders.

6. Experimental laboratory tests

The purpose of these experiments is to test the concepts of the super absorbing air filters and DATD designed and carried out by the authors, previously presented in detail. Testing was performed on an experimental stand, located in the Laboratory of Internal Combustion Engines of the Faculty of Engineering of Hunedoara, Romania.

The data were processed and compared with those obtained when operation took place the original engine filter provided by the manufacturer. There is clear improvement of pressure on the inlet route, when super absorbing filters and the dynamic air transfer device are installed.

The experimental measurements were based on a stand containing a 4-stroke 4-vertical inline cylinder spark ignition engine, the camshaft in the crankcase, Dacia brand, model

810.99, with carburettor, and related equipment, stand which allows setting the pressure field on the engine intake route (Figure 44), for example see [2,3].

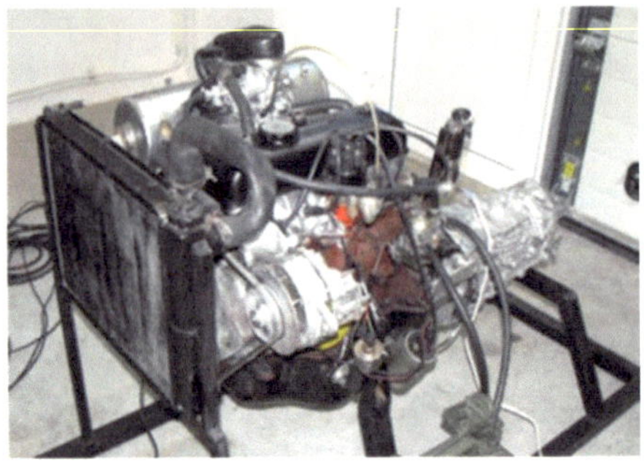

Figure 44. Overview of experimental stand

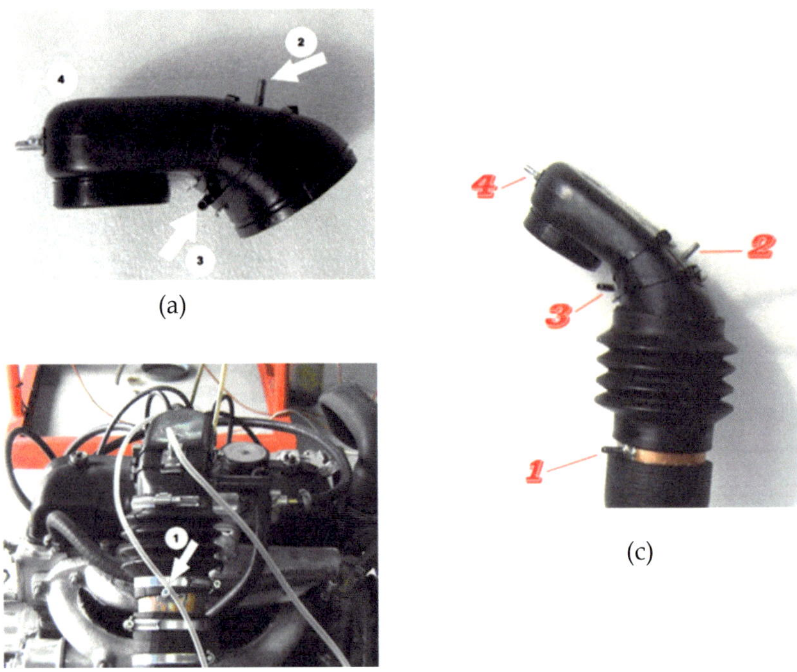

(a) (b) (c)

Figure 45. a, b, c.; Position and number of pressure intake ports

A number of the pressure intake ports were made downstream the air filter and measurements were made at different operating regimes for the engine installed on the stand, for different super absorbing filters designed and made by the authors. The position of the pressure intake ports on the engine intake route is illustrated in Figure 45.

Measurements were performed in no-load (idling) engine motion at various revolutions. Relative pressure values were measured on the intake route points where pressure ports have been mounted, as shown in Figure 45. TESTO 510 digital manometer (0-100hPa) was used. For example see [3].

Also, to simulate vehicle movement, measurements were performed in the presence of a fan positioned in front of the cooling radiator of the engine installed on the stand.

Measuring the speed of airflow from the fan to the engine radiator took place using a digital anemometer, Lutron LM - 8010 type. Engine revolution was measured with a VELLEMAN DTO 6234N digital tachometer.

In addition, engine noise measurements were made for operation with different filter types, with a Lutron SL - 4012 type sound level meter.

Data were collected for the inlet system equipped with original classic air filter – OAF (Figure 46), super absorbing cylindrical air filter with internal diffuser – SAAFid (Figure 47), supliform super absorbing air filter – suSAAF (Figure 48), super absorbing filter with wide filtration range – SAAFwr (Figure 49) and for the dynamic air transfer device DATD (figure 50).

Figure 46. Original classic filter trial

Figure 47. Super absorbing cylindrical filter with internal diffuser trial

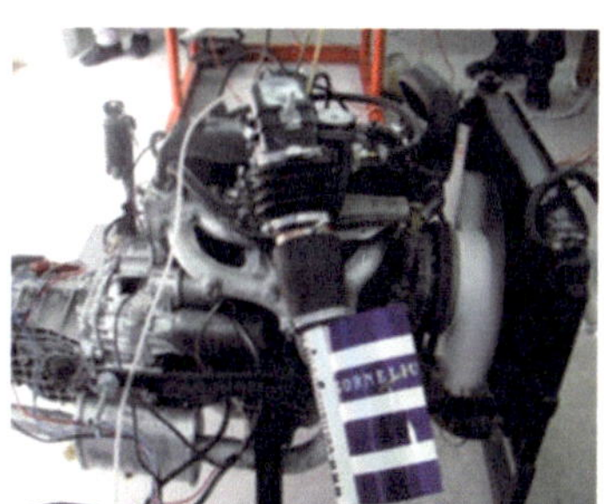

Figure 48. Supliform super absorbing filter trial

Figure 49. Super absorbing with wide filtration range trial

(a) (b) (c)

Figure 50. a, b, c: Dynamic air transfer device DATD trial

Further are presented comparative graphs of the relative pressure values recorded for each concept, for each individual pressure intake port.

Due to the presence of pressure waves generated by alternative movement of the pistons in the cylinders and the periodic opening and closing of the intake valves, pressure values fluctuate within a fairly wide range. Therefore, after stabilization of engine revolution, limit values (upper and lower) of the pressure in the ports mounted were registered and their average was calculated. These averages were used for plotting the graphs below.

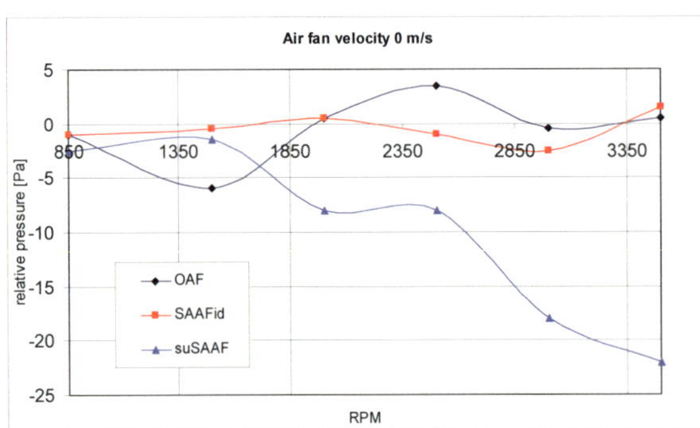

OAF – Original Air Filter, SAAFid – Super Absorbing Air Filter with internal diffuser, suSAAF – supliform Super Absorbing Air Filter

Figure 51. Values for pressure intake port 1, without vehicle movement simulation

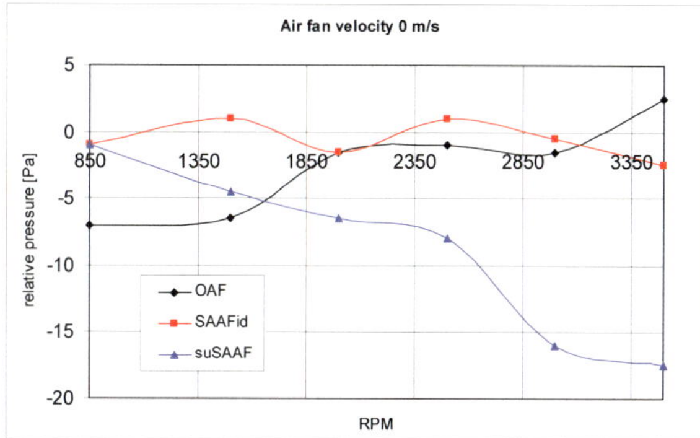

OAF – Original Air Filter, SAAFid – Super Absorbing Air Filter with internal diffuser, suSAAF – supliform Super Absorbing Air Filter

Figure 52. Values for pressure intake port 2, without vehicle movement simulation

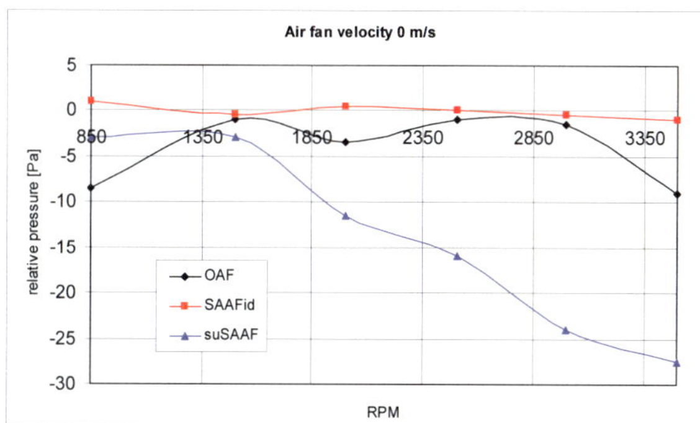

OAF – Original Air Filter, SAAFid – Super Absorbing Air Filter with internal diffuser, suSAAF – supliform Super Absorbing Air Filter

Figure 53. Values for pressure intake port 3, without vehicle movement simulation

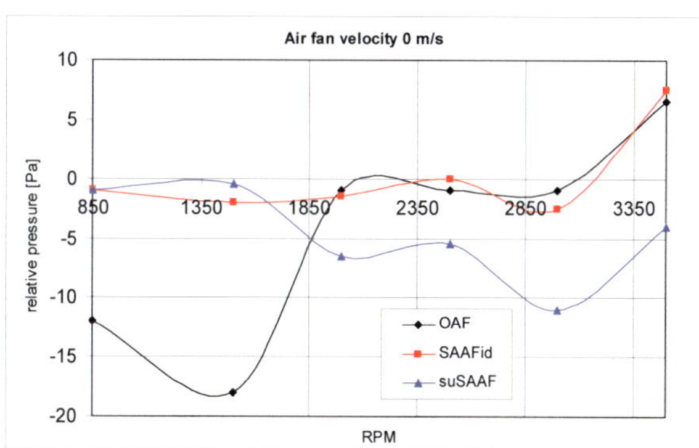

OAF – Original Air Filter, SAAFid – Super Absorbing Air Filter with internal diffuser, suSAAF – supliform Super Absorbing Air Filter

Figure 54. Values for pressure intake port 4, without vehicle movement simulation

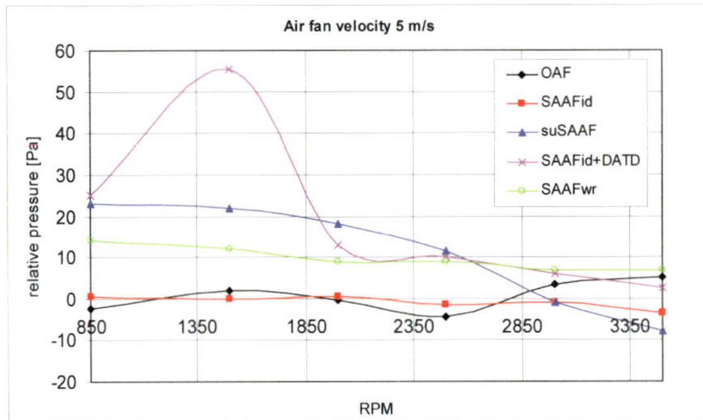

OAF – Original Air Filter, SAAFid – Super Absorbing Air Filter with internal diffuser, suSAAF – supliform Super Absorbing Air Filter, DATD – Dynamic Air Transfer Device, SAAFwr – Super Absorbing Air Filter with wide filtration range

Figure 55. Values for pressure intake port 1, with vehicle movement simulation

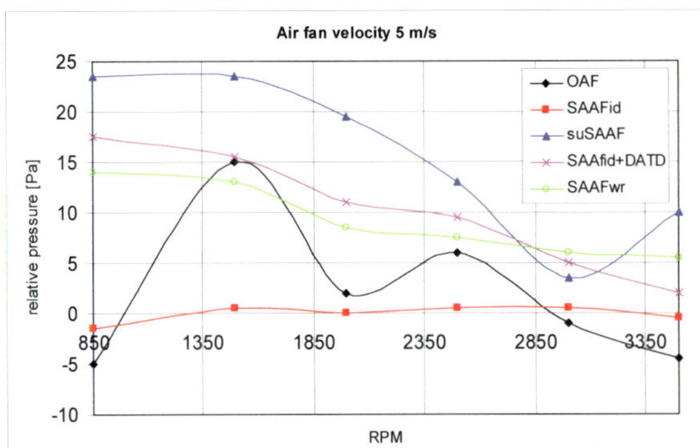

OAF – Original Air Filter, SAAFid – Super Absorbing Air Filter with internal diffuser, suSAAF – supliform Super Absorbing Air Filter, DATD – Dynamic Air Transfer Device, SAAFwr – Super Absorbing Air Filter with wide filtration range

Figure 56. Values for pressure intake port 2, with vehicle movement simulation

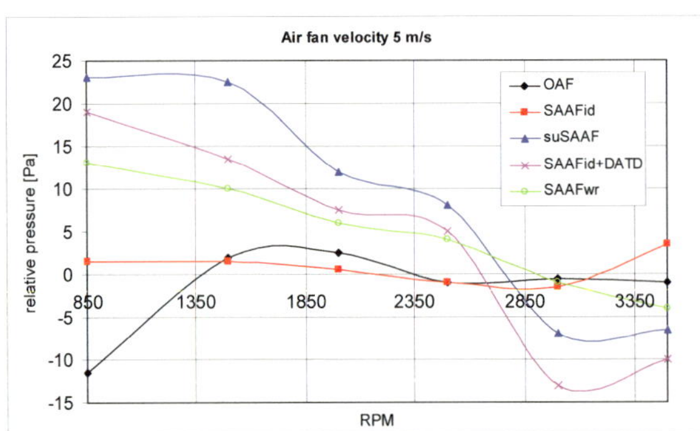

OAF – Original Air Filter, SAAFid – Super Absorbing Air Filter with internal diffuser, suSAAF – supliform Super Absorbing Air Filter, DATD – Dynamic Air Transfer Device, SAAFwr – Super Absorbing Air Filter with wide filtration range

Figure 57. Values for pressure intake port 3, with vehicle movement simulation

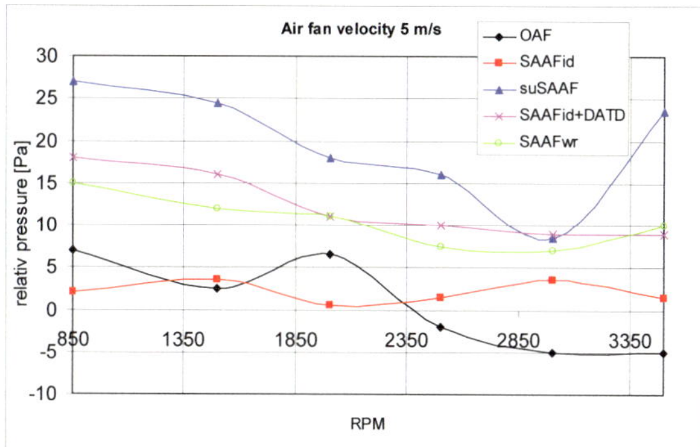

OAF – Original Air Filter, SAAFid – Super Absorbing Air Filter with internal diffuser, suSAAF – supliform Super Absorbing Air Filter, DATD – Dynamic Air Transfer Device, SAAFwr – Super Absorbing Air Filter with wide filtration range

Figure 58. Values for pressure intake port 4, with vehicle movement simulation

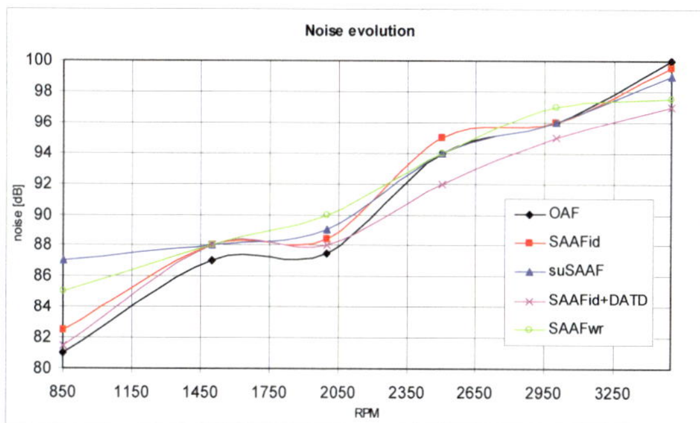

OAF – Original Air Filter, SAAFid – Super Absorbing Air Filter with internal diffuser, suSAAF – supliform Super Absorbing Air Filter, DATD – Dynamic Air Transfer Device, SAAFwr – Super Absorbing Air Filter with wide filtration range

Figure 59. Evolution of noise depending on engine revolution

The following are comparative graphs of the evolution of relative pressure on the intake route for each revolution regime, with vehicle movement simulation (air fan velocity 5 m/s).

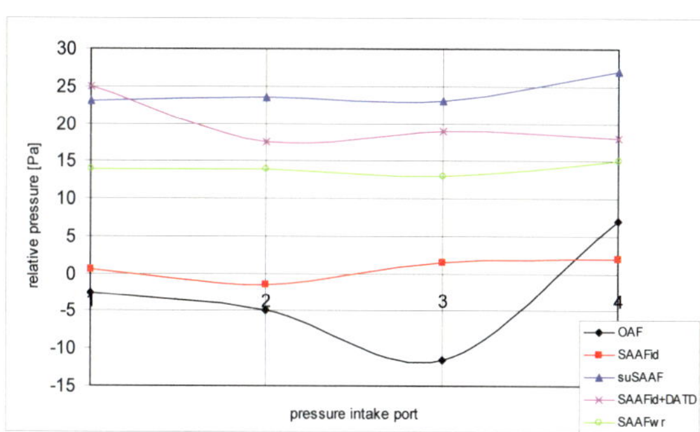

OAF – Original Air Filter, SAAFid – Super Absorbing Air Filter with internal diffuser, suSAAF – supliform Super Absorbing Air Filter, DATD – Dynamic Air Transfer Device, SAAFwr – Super Absorbing Air Filter with wide filtration range

Figure 60. Evolution of relative pressure on intake route, 800 RPM

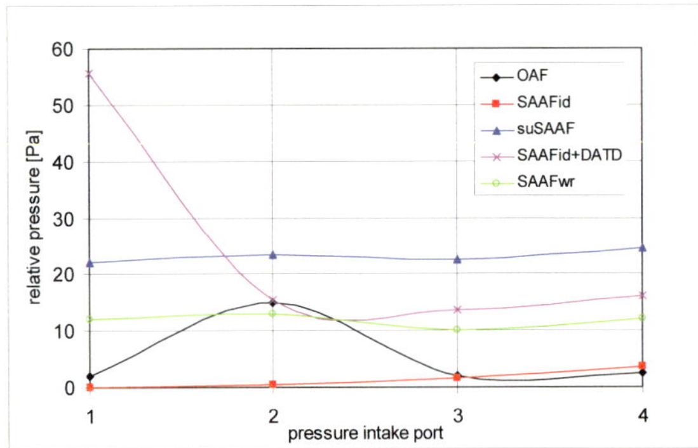

OAF – Original Air Filter, SAAFid – Super Absorbing Air Filter with internal diffuser, suSAAF – supliform Super Absorbing Air Filter, DATD – Dynamic Air Transfer Device, SAAFwr – Super Absorbing Air Filter with wide filtration range

Figure 61. Evolution of relative pressure on intake route, 1500 RPM

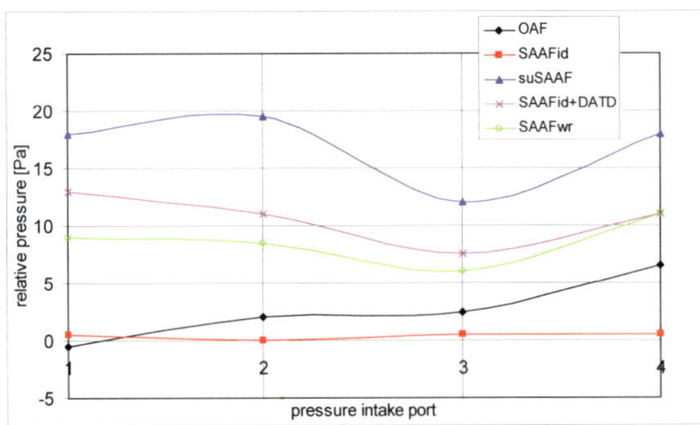

OAF – Original Air Filter, SAAFid – Super Absorbing Air Filter with internal diffuser, suSAAF – supliform Super Absorbing Air Filter, DATD – Dynamic Air Transfer Device, SAAFwr – Super Absorbing Air Filter with wide filtration range

Figure 62. Evolution of relative pressure on intake route, 2000 RPM

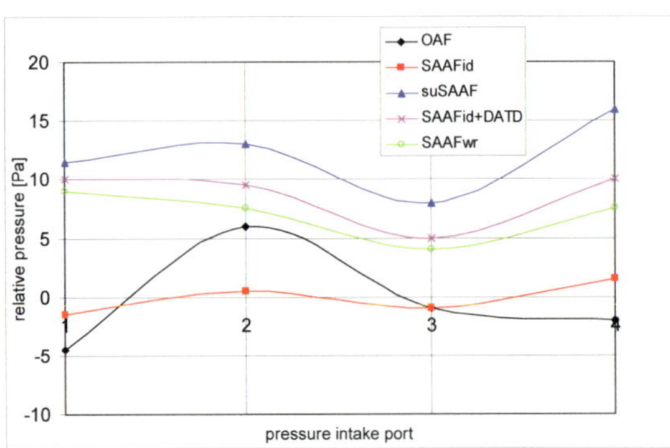

OAF – Original Air Filter, SAAFid – Super Absorbing Air Filter with internal diffuser, suSAAF – supliform Super Absorbing Air Filter, DATD – Dynamic Air Transfer Device, SAAFwr – Super Absorbing Air Filter with wide filtration range

Figure 63. Evolution of relative pressure on intake route, 2500 RPM

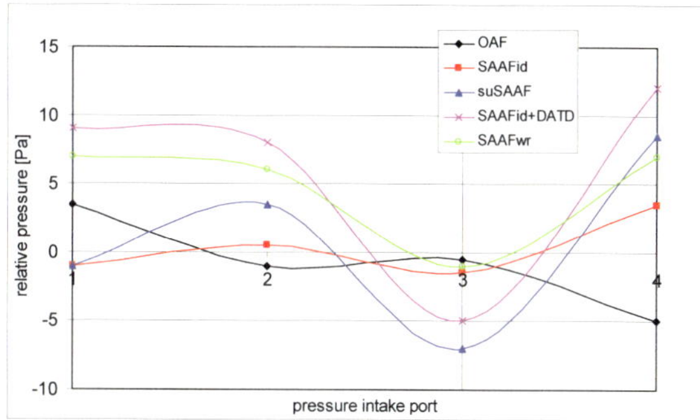

OAF – Original Air Filter, SAAFid – Super Absorbing Air Filter with internal diffuser, suSAAF – supliform Super Absorbing Air Filter, DATD – Dynamic Air Transfer Device, SAAFwr – Super Absorbing Air Filter with wide filtration range

Figure 64. Evolution of relative pressure on intake route, 3000 RPM

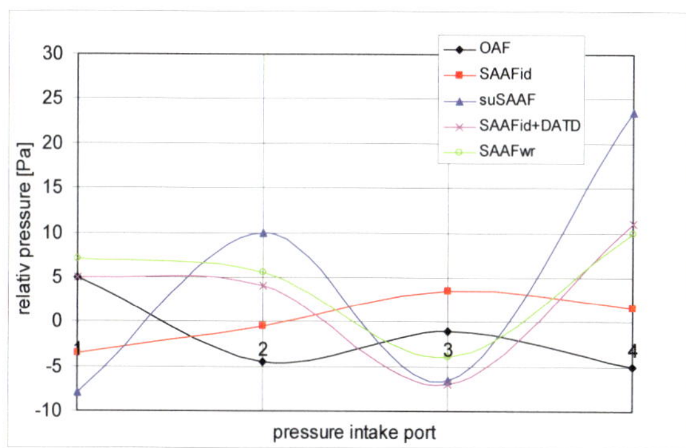

OAF – Original Air Filter, SAAFid – Super Absorbing Air Filter with internal diffuser, suSAAF – supliform Super Absorbing Air Filter, DATD – Dynamic Air Transfer Device, SAAFwr – Super Absorbing Air Filter with wide filtration range

Figure 65. Evolution of relative pressure on intake route, 3500 RPM

7. Conclusions

Studying the pressure evolution for each pressure intake port and at different engine revolutions (RPM), we can conclude the following [2,4]:

1. The appearance of a slight superpressure effect introduced by the super absorbing filters with internal diffuser (SAAFid) and with wide filtration range (SAAFwr), throughout the revolution range, which is beneficial to the filling process;
2. The superpressure effect is more noticeable at low revolutions, for all filter concepts and decreases with revolution increase;
3. The super absorbing filters with internal diffuser and with wide filtration range retain the superpressure effect for high revolutions;
4. The fluctuation of pressure depending on revolution is less acute in super absorbing filters with internal diffuser and with wide filtration range;

5. Following the evolution of pressure on the intake route for different engine revolutions it is found that the filters designed have much diminished pressure fluctuation as compared to the one induced by the original classic air filter (OAF) for low revolutions. At high revolutions, the evolution of pressure is somewhat similar for all filters.

Seen as a whole, we can say that the super absorbing air filters (SAAF) together with the dynamic air transfer device (DATD) and the integrated deflector (ID) for attenuation of thermal radiation coming from the cooling radiator, lead to the following advantages:

- Reducing pressure losses on the air intake route in the engine cylinders, which leads to an increase in the inlet air pressure and thus to an increase in the amount of air retained in the cylinders during an engine cycle;
- Less acute heating of the air filter during engine operation, especially in summer. The low air filter temperature means less intense heating of the filtered air during engine operation, with beneficial effects on air density. Cooler air means higher density, i.e. larger amount of air retained in the engine cylinders during an engine cycle.

Author details

Sorin Raţiu[1,2,*] and Corneliu Birtok-Băneasă[1,3,4]
[1]*"Politehnica" University of Timisoara, Romania,*
[2]*Engineering Faculty of Hunedoara, Romania,*
[3]*Mechanical Faculty of Timisoara, Romania,*
[4]*Corneliu Group, Romania*

8. References

[1] Raţiu S, Mihon L (2008) *Internal Combustion Engines for Motor Vehicles - Processes and Features*, MIRTON Publishing House, Timişoara, pp. 44-47;

[2] Birtok-Băneasă C, Raţiu S (2011) *Air Intake of Internal Combustion Engines – Super Absorbing Filters - Dynamic Transfer Devices*, POLITEHNICA Publishing House, Timişoara, pp. 15-90;

[3] Raţiu S (2009) *Internal Combustion Engines for Motor Vehicles - Processes and Features – Laboratory Experiments*, MIRTON Publishing House, Timişoara, pp. 40-42;

[4] Raţiu S, Birtok-Băneasă C, Alic C, Mihon L (2009) *New concepts in modeling air filters for internal combustion engines*, 20th International DAAAM SYMPOSIUM "Intelligent Manufacturing & Automation: Theory, Practice & Education", Vienna, Austria, ISSN 1726-9679

[5] www.corneliugroup.ro

* Corresponding Author

Chapter 9

Understanding Fuel Consumption/Economy of Passenger Vehicles in the Real World

Yuki Kudoh

Additional information is available at the end of the chapter

http://dx.doi.org/10.5772/49944

1. Introduction

The world is currently highly dependent upon oil for automotive transport. As a result, large amounts of greenhouse gas (GHG) emissions are generated in the passenger automotive sector and are having a substantial effect on the environment. Among the various measures from both the automotive technology side and the transport demand side to reduce energy consumption and GHG emissions, improving the fuel consumption (FC, expressed in litres of gasoline per hundred kilometres of travel [L/100 km]) or fuel economy (FE, usually expressed in [km/L] or miles per gallon [mpg]) of passenger vehicles is regarded as the most effective measure. In this regard, many regions and countries around the world have implemented FC/FE or GHG standards (An et al., 2011), and some – for example, the United States, the European Union, and Japan – are tightening their existing standards.

FC/FE and GHG standards are measured by using chassis dynamometer test cycles, which simulate a variety of driving conditions at typical highway and urban driving speeds in each country and region. However, it is quite well known that a gap exists between FC/FE values generated by dynamometer testing and real-world values worldwide (Schipper & Tax, 1994; Schipper, 2011). Real-world FC/FE values depend strongly upon each driver's style and location, upon the traffic congestion, weather, and corresponding use of accessories (especially air conditioning), and upon the vehicle's maintenance condition. No single test cycle can simulate all possible combinations of these factors. Although the energy roadmaps and CO2 reduction targets for the passenger automotive sector in each region and country are based mainly on FC/FE and GHG standards, it is the real-world values that matter. It is doubtful whether reduction targets can be met without more accurate real-world assessments of FC/FE and GHG emissions. Providing more accurate real-world FC/FE values will also better inform consumers of expected fuel costs.

The many studies that have investigated FC/FE and GHG emissions in the real world have used several approaches. Schipper (2011) analysed FC trends for the entire passenger vehicle fleet in the United States, Australia, Japan, and several European countries by using national or regional statistics, as well as the impact of fuel prices upon FC/FE. Wang et al. (2008) explored the influence of driving patterns on FC in China by using a portable emissions measurement system and established an on-road FC estimation model. Duoba et al. (2005) tested the robustness of FE to changes in vehicle activity for hybrid vehicles (HVs) and their counterpart internal combustion engine vehicles by applying various driving schedules upon a chassis dynamometer in the U.S. Several studies have also analysed on-road FC/FE by using information collected by questionnaires or on the internet. Huo et al. (2011) examined the differences between standard test and real-world values for Chinese passenger vehicles by using data voluntarily reported by drivers on the internet; the study gathered 63,115 pieces of real-world FC data for 153 vehicle models. Sagawa & Sakaguchi (2000) analysed the FE of Japanese passenger vehicles using questionnaires, but they were unable to analyse the data with high statistical reliability because of sample number limitations (1,479 samples).

The use of internet-connected mobile phones has become widespread throughout the world, and the range of mobile phone contents and services provided includes those used to track the FC/FE of automobiles. In order to analyse real-world FC/FE in Japan with a high level of statistical reliability, the author's group put focus upon the FC/FE management service in which voluntarily reported FC log data of the vehicle users are collected through internet-connected mobile phones across Japan and developed and on-road (actual) FC database. The findings for the 24 months from October 2000 to September 2002 are reported by Kudoh et al. (2004).

Since 2000, the number of brand-new passenger vehicles sold in Japan has fluctuated between 4.26 and 4.76 million, which means that about 8% of the total passenger vehicle fleet was replaced with brand-new vehicles annually. Since the FC/FE of brand-new vehicles improved during this period, it is reasonable to conclude that the FC/FE performance of the passenger vehicle fleet itself should also have improved as these new vehicles replaced older ones. In addition, the mobile phone service provider whose log data were used to develop the actual FC database has reported an increase in the number of users in the 2000s.

The author's group therefore updated the database by extending the data collection period from 24 to 54 months and created a database consisting of 1,645,923 pieces of log data collected from October 2000 through March 2005, including information from 49,677 passenger vehicle users on 2,022 models sold in Japan and conducted a statistical analysis of actual FC for passenger HVs and other passenger vehicles with internal combustion engine in Japan (Kudoh et al., 2007; Kudoh et al., 2008). In addition to the previous achievements of the author's group, this paper addresses the effects of vehicle specifications towards the actual FC/FE of passenger vehicles in Japan, as derived from the database, from a statistical point of view.

2. Reasons for the FC/FE measurement gap

In Japan, targets for FE standards are provided in the revised Law Concerning the Rational Use of Energy (known as the Energy-Saving Law) by implementing the "Top Runners

Approach," which aims to establish energy-efficiency standards that meet or exceed the best energy-efficiency specifications for a product in an industry.

According to the revised law, passenger vehicles sold on the Japanese market in 2010 were expected to achieve the FE standard stipulated in the Japanese 10-15 mode driving schedule for each vehicle inertia weight class. The 10-15 mode driving schedule was developed for exhaust measurement and FE tests of light duty vehicles in Japan, including passenger vehicles; the driving pattern and relationship between velocity and acceleration are shown in Figure 1. The test is conducted on a chassis dynamometer with a hot start at curb weight plus 110 [kg] (the approximate weight of 2 passengers), with the air conditioner and other electrical appliances turned off. Figure 2 shows an example of actual vehicle travel activity measured in an urban area (TMGBE, 1996); the average velocity is almost the same in both figures. Although the 10-15 mode driving schedule is supposed to represent actual vehicle travel activity within Japanese urban areas, acceleration and deceleration occurred more frequently under actual conditions and higher levels of acceleration were observed at low velocities. These factors are thought to be among the main reasons for the gap between 10-15 mode FE and actual FE values.

Figure 3 depicts the simulated results of the 10-15 mode FC and the actual FC for a passenger gasoline engine vehicle (GV) with a 2,000 cc displacement. The results were calculated under vehicle driving simulation model (Kudoh et al., 2001). At a similar average velocity (as shown in Figure 2), the actual FC was about 13% lower than predicted by the 10-15 mode test. In addition, the actual FC of a vehicle clearly varied according to where it was driven, because the main cause of changes in average velocity is the stop-and-go traffic pattern that occurs frequently in urban areas.

As pointed out by Farrington & Rugh (2000) and Nishio et al. (2008), another important factor that should affect the FC/FE gap is the use of air conditioning, because the air conditioning system is turned off in most test cycles on the chassis dynamometer, including in the Japanese 10-15 mode.

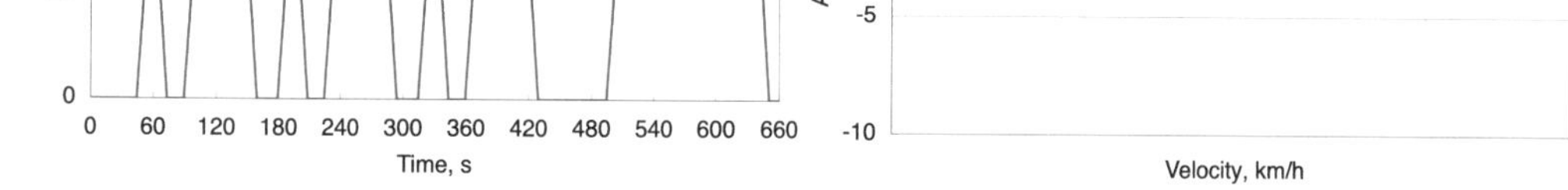

(a) Driving pattern

(b) Relationship between velocity and acceleration

Figure 1. Japanese 10-15 mode driving schedule.

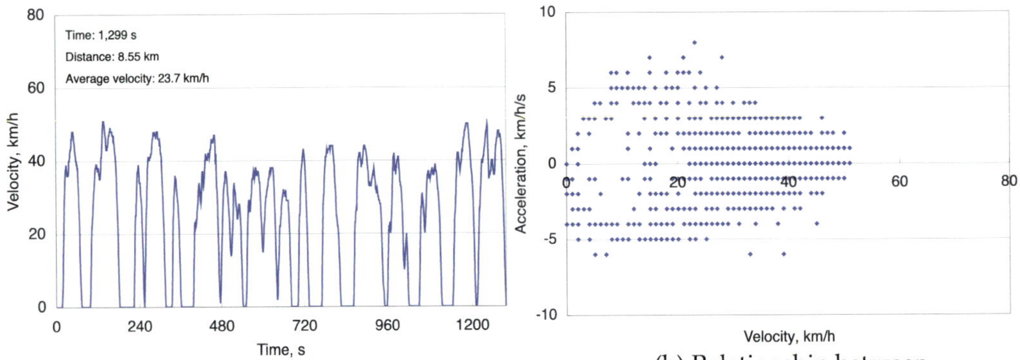

(a) Driving pattern

(b) Relationship between velocity and acceleration

Figure 2. An example of actual vehicle travel activity (TMGBE, 1996).

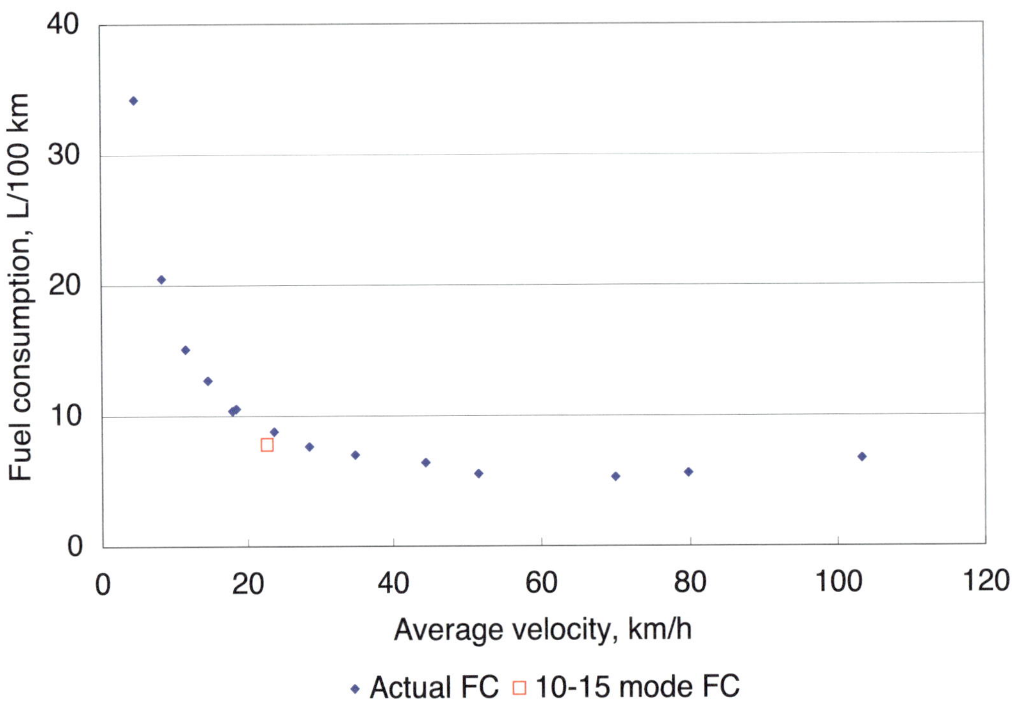

Figure 3. Simulated FC of a passenger vehicle with a 2,000cc gasoline engine.

3. Outline of the actual FC database

Figure 4 outlines the actual FC database that the author's group has been developing based upon the catalogue data of passenger vehicles sold in Japanese market and the voluntarily reported FC log data of vehicle users.

To obtain the passenger vehicle specifications for cars sold in Japan before March 2005, vehicle catalogues for each vehicle name, model year, and model grade were downloaded from an available website on the internet. The vehicle specification database contained information on 35,177 vehicles.

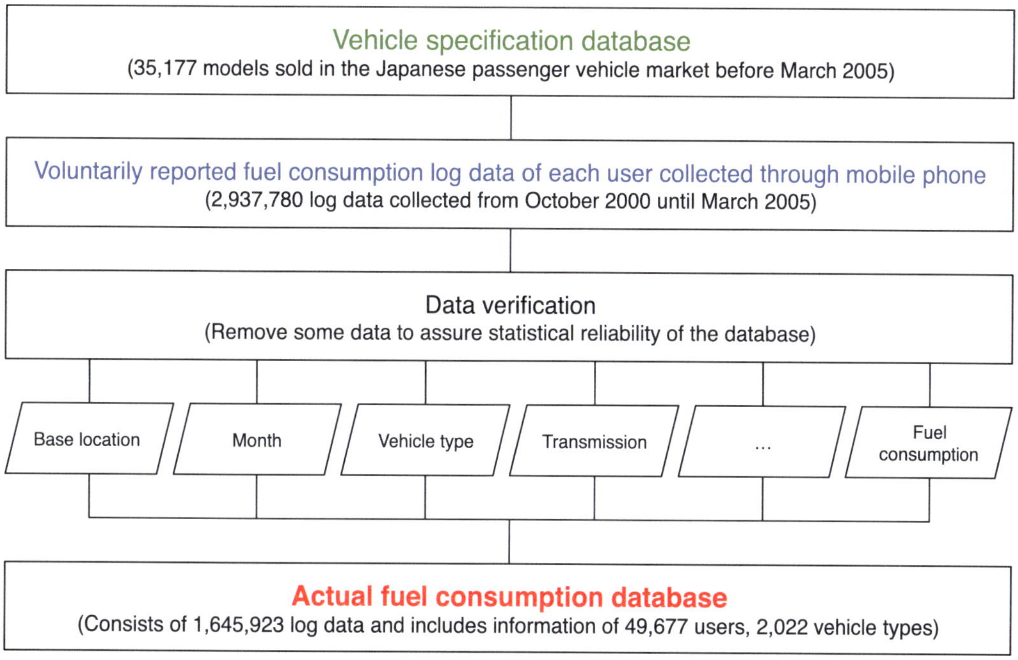

Figure 4. Outline of the actual FC database.

The actual FC database was developed by using voluntarily reported FC data from vehicle users and the vehicle specification database. The FC data collection system is called *e-nenpi* (which stands for "electronic FE" in Japanese[1]); this is an online service for internet-connected mobile phone users[2] provided by IID, Inc. The system manages information for vehicle owners, including FC performance and recommended routine maintenance. Users of the service register and provide the following information: (1) zip code of residence, (2) vehicle type[3], (3) type of engine air intake (turbocharged/supercharged or normal), (4) transmission type (manual or automatic[4]), and (5) type of fuel used (unleaded gasoline, premium unleaded gasoline, diesel, or liquefied petroleum gas). Through their mobile phone, the user then enters the amount of fuel put into the vehicle's tank and the odometer reading at the time of fuelling, and the user's FC data are stored on a server.

The items required for service registration were linked with the vehicle specification database and supplemented with other items such that the following 17 attributes were

---

[1] More information is available at: http://e-nenpi.com (in Japanese).

[2] Although the service was originally provided only for internet-connected mobile phone users, the provider currently offers the service for personal computers as well.

[3] Vehicle type is a code prepared by vehicle makers and approved by the government for vehicles sold and used in Japan to identify each vehicle.

[4] Although the FC may differ depending on the type of automatic transmission, they are grouped together within the database owing to data restrictions.

included in the actual FC database for each user: (1) user ID, (2) base location of where the vehicle was used[5], (3) month and year when the vehicle was fuelled, (4) vehicle maker, (5) vehicle name, (6) vehicle type, (7) vehicle class (light passenger vehicle[6] (LP) or passenger vehicle (P)), (8) type of powertrain (gasoline vehicle (GV), diesel vehicle (DV) or hybrid vehicle (HV)), (9) type of air intake, (10) transmission type, (11) type of drive system (2WD or 4WD), (12) type of fuel injection engine (direct injection or not), (13) whether a variable valve timing system was used, (14) fuel tank capacity, (15) engine displacement, (16) vehicle kerb weight, and (17) 10-15 mode FE.

Although technological specifications may vary within the same vehicle type by grade or model year owing to differences in equipment or improvement in vehicle technologies, the model year of the vehicle owned by each user could not be specified from the log data. Hence, the following values obtained from the vehicle specification database were used in the technological specifications of a vehicle type in the actual FC database: (1) maximum fuel tank capacity, (2) simple average of minimum and maximum vehicle weight, and (3) simple average of minimum and maximum 10-15 mode FE.

A total of 2,937,780 FC log data points was collected over the 54-month study period (from October 2000 through March 2005). Data were excluded under the following conditions to assure the statistical reliability of the database:

a. when the base location of vehicle use could not be specified (21,736 entries);
b. when users specified a vehicle type that was not included in the vehicle specification database (611,357 entries); and
c. when the fuel fill-up rate (γ) was less than 60% or more than 100% (536,620 entries). The rate was calculated as $\gamma = f / C$, where f [L] is the amount of fuel put into the tank and C [L] is the fuel tank capacity.

$FE_{u,v}$ [km/L], the FE of user u who owns vehicle type v, was calculated by Equation 1, where $d_{u,v,i}$ [km] is driving distance from the last fuelling of the i th data point, $f_{u,v,i}$ [L] is the amount of fuel obtained for data point i, and $n_{i\in u,v}$ is the number of log data entries.

$$FE_{u,v} = \sum\nolimits_{i\in u,v} (d_{u,v,i} / f_{u,v,i}) / n_{i\in u,v} \quad (1)$$

FE_v [km/L], the FE of vehicle type v, was calculated by using Equation 2, where $n_{u\in v}$ is the number of users who own v.

$$FE_v = \sum\nolimits_{u\in v} FE_{u,v} / n_{u\in v} \quad (2)$$

Data entries were eliminated from further analysis if they met any of the following conditions:

---

[5] This was determined from the zip code provided by the registered user.

[6] A light passenger vehicle is equivalent to, or smaller than, the EU's A-segment. Its physical size and engine power are regulated as follows: maximum length, 3.39 [m]; maximum width, 1.48 [m]; maximum height, 2 [m]; maximum engine displacement, 660 [cc]; and maximum engine power, 64 [hp].

a. when $FE_{i \in u,v}$ was determined to be a statistical outlier by the Grubbs' test at a critical level of 5% (57,118 entries);
b. when $n_{i \in u,v}$ is less than 5 (10,773 entries);
c. when the variance of $FE_{u \in v}$ is greater than 10 [(km/L)^2] (4,789 entries);
d. when $FE_{u \in v}$ was determined to be a statistical outlier by the Grubbs' test at a critical level of 5% (10,414 entries); and
e. when $n_{u \in v}$ is less than 3 (40,050 entries).

After all of the eliminations, 1,645,923 log data points, including pieces of information from 49,677 users and 2,022 vehicle types, were used to develop the actual FC database. A summary of the number of data points, users, and vehicle types is given in Table 1.

Vehicle type		Number of log data points	Number of users	Number of vehicle types
Light passenger gasoline vehicle (LP-GV)	< 702 kg	23,563	848	57
	703 – 827 kg	70,745	2,189	112
	828 – 1,015 kg	55,655	1,654	93
	1,016 – 1,265 kg	1,779	43	3
	Total	151,742	4,734 (0.035%[1])	265 (54.6%[2])
Passenger diesel vehicle (P-DV)	1,016 – 1,265 kg	91	4	1
	1,266 – 1,515 kg	496	19	5
	1,516 – 1,765 kg	8,040	236	27
	1,766 – 2,015 kg	28,021	809	57
	2,016 – 2,265 kg	18,159	477	22
	2,266 kg +	188	7	1
	Total	54,995	1,552 (0.061%[1])	113 (21.4%[2])
Passenger gasoline vehicle (P-GV)	< 702 kg	1,179	48	4
	703 – 827 kg	10,626	380	20
	828 – 1,015 kg	120,105	4,005	169
	1,016 – 1,265 kg	346,834	10,968	481
	1,266 – 1,515 kg	600,790	17,468	567
	1,516 – 1,765 kg	281,517	8150	285
	1,766 – 2,015 kg	51,055	1543	84
	2,016 – 2,265 kg	18,398	526	24
	2,266 kg +	3298	87	2
	Total	143,3802	43,175 (0.108%[1])	1,636 (41.3%[2])
Passenger (gasoline) hybrid vehicle (P-HV)	703 – 827 kg	66	4	1
	828 – 1,015 kg	379	12	1
	1,016 – 1,265 kg	2,455	86	3

Vehicle type		Number of log data points	Number of users	Number of vehicle types
	1,266 – 1,515 kg	671	43	1
	1,766 – 2,015 kg	1,447	51	1
	2,016 – 2,265 kg	366	20	1
	Total	5,384	216 (0.111%[1])	8 (57.1%[2])
Total		1,645,923	49,677 (0.089%[1])	2,022 (40.5%[2])

Table 1. Data size categories of the actual FC database. The vehicle weight class follows the Japanese inertia weight classes for passenger vehicles. [1] Sampling rate relative to the number of vehicles owned as of March 2005. [2] Sampling rate relative to the number of vehicle types included in the vehicle specification database.

Although Equations 1 and 2 assume that users fill their tanks to the same (full) level at every refuelling, there may be users who do not do so. The *e-nenpi* system recommends that registered users fill up the vehicle tank, and a confirmation message to check whether they have filled up the tank is shown when they input their fuel log through the mobile phone. The second and subsequent log data entries were saved in the server only after a user had confirmed filling up more than twice. In addition, some data were eliminated if they did not satisfy criterion c; the average of fuel fill-up rate of the remaining log data was 76.8% (standard deviation = 8.82%). Users should refuel before the tank was completely empty, indicating that most of the user data included in the actual FC database were acquired as the users filled up at petrol stations, so the fill level of the vehicles was expected to be almost the same every time.

4. Vehicle specifications and actual FC/FE

In the Japanese passenger vehicle market, 12 HV types had been launched as of March 2005; 8 were included in the actual FC database. It is assumed that the FC/FE performance of these vehicles would vary with differences in the powertrain configuration (e.g., series hybrid, parallel hybrid, or power-split hybrid) or degree of hybridisation (such as full hybrid, power-assist hybrid, mild hybrid, or plug-in hybrid). However, because of the difficulties involved in including all of these factors with a high level of statistical reliability, the passenger HV types were combined in this study.

4.1. Japanese 10-15 mode and actual FE

Figure 5 depicts the relationship between the Japanese 10-15 mode FE and actual FE. $FE_{v,actual}$ [km/L], the actual FE of vehicle type v, was calculated from Equation 3 (USEPA 2010), where $d_{v,i}$ and $f_{v,i}$ are driving distance [km] and amount of fuel [L] at i th log data point of vehicle v.

$$FE_{v,actual} = \sum_{i \in v} d_{v,i} / \sum_{i \in v} f_{v,i} \quad (3)$$

Table 2 shows the results of a linear regression analysis and the 95% confidential interval (95 CI) described by Equation 4, where $FE_{v,10\text{-}15}$ [km/L] is the 10-15 mode FE of vehicle v .

$$FE_{v,actual} = a \cdot FE_{v,10-15} \tag{4}$$

If a plotted point was on the diagonal line shown in Figure 5, the actual FE of the vehicle was exactly the same as 10-15 mode FE. As can be seen in the figure, the gap between 10-15 mode FE and actual FE increased as the 10-15 mode FE increased.

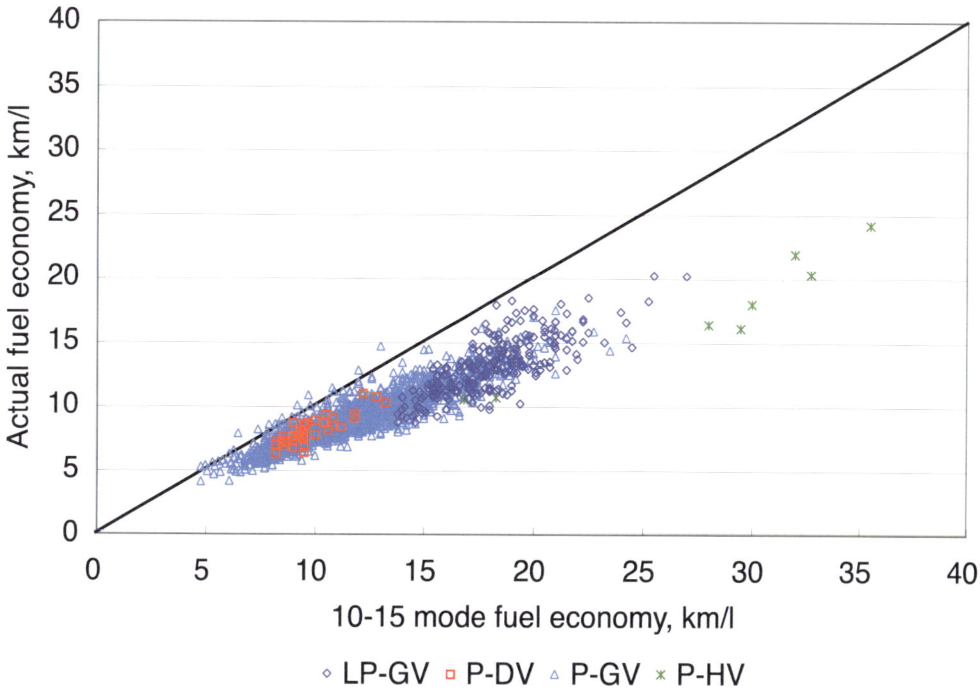

Figure 5. 10-15 mode FE and actual FE.

	LP-GV	P-DV	P-GV	P-HV
n	240	36	1,352	8
R^2	0.989	0.995	0.989	0.994
a (95 CI)	0.725 (0.715–0.735)	0.823 (0.803–0.844)	0.760 (0.756–0.765)	0.622 (0.579–0.666)
t	144.3	82.6	346.1	34.0

Table 2. Estimates of parameters by Equation 4. t is the t statistics.

25 P-GVs had an actual FE that was higher than the corresponding 10-15 mode FE. (These are above the line in Figure 5.) Figure 6 shows the achievement ratio of actual FE to 10-15 mode FE of domestically produced and imported P-GVs; 23 out of the 25 P-GVs with a ratio of greater than 1 were imported vehicles. These results indicate that the achievement ratio of actual FE to 10-15 mode FE may be higher for imported vehicles than for domestically produced vehicles. The results of a two-tailed Welch test confirmed that the mean

achievement ratios of domestically produced passenger vehicles (x) were significantly lower than those of imported passenger vehicles (y) (mean of $x = 0.758$, variance of $x = 0.00445$, mean of $y = 0.854$, variance of $y = 0.00810$, $T = 15.1$, degree of freedom = 268; $p < 0.05$). One possible explanation is that the drivetrains or transmissions of imported vehicles are not optimised for Japanese road conditions and their 10-15 mode FEs tend to be lower than their counterpart domestically produced P-GVs.

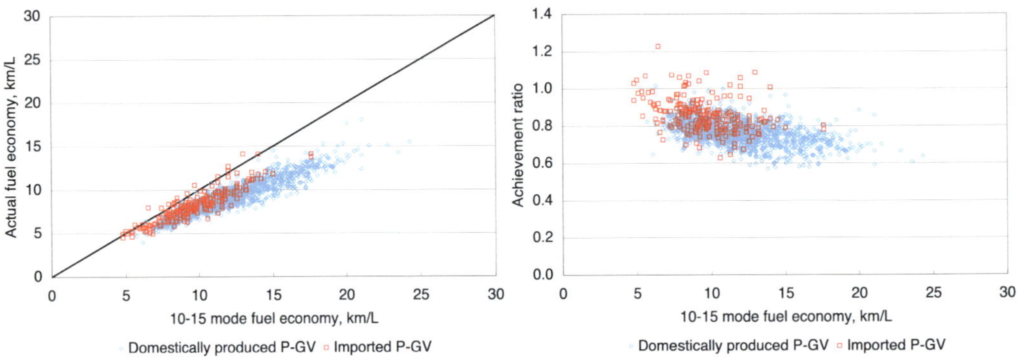

(a) 10-15 mode FE and actual FE

(b) Achievement ratio of actual FE to 10-15 mode FE

Figure 6. Comparison of domestically produced P-GVs and imported P-GVs.

4.2. Vehicle weight and actual FC

Weight-saving technologies in passenger vehicles will play an important role in improving FC, along with improvements in engine and drivetrain efficiency. Figure 7 depicts the relationship between vehicle weight and actual FC. Here, $FC_{v,actual}$ [L/100 km], the actual FC of vehicle type v, is calculated by Equation 5.

$$FC_{v,actual} = 100 \cdot \sum_{i \in v} f_{v,i} / \sum_{i \in v} d_{v,i} \tag{5}$$

Two FC standards are shown in Figure 7: the Japanese 2010 standard for GVs and the 2005 standard for DVs. Points plotted above the two lines represent vehicles that do not achieve the FC standards in the real world. Although most brand-new passenger vehicles were announced to have achieved the FC standard by 2005, Figure 7 reveals that only some P-DVs and all the P-HVs achieved the Japanese FC standard in the real world at that time.

Since $FC_{v,actual}$ can be thought to be proportional to vehicle weight w [kg], a linear regression analysis was conducted by using Equation 6 (Table 3).

$$FC_{v,actual} = b \cdot w + c \tag{6}$$

The analysis showed that it is difficult to explain the FC of LP-GVs and P-DVs only by vehicle weight. Sales of brand-new LP-GVs, which are restricted in terms of vehicle size and engine displacement, are rapidly expanding in Japan, and Japanese vehicle makers provide a variety of vehicle types (e.g., hatchbacks and wagons) within the regulatory standard. To compensate for the increase in vehicle weight incurred by equipment installed to meet consumer needs or to satisfy safety standards, many LP-GVs use turbochargers. Including only LP-GVs that were introduced to the market after 1998 (when the LP vehicle standards were changed to meet new crash safety standards), the engine displacement of LP-GVs were from 657 – 660 [cc] but their average vehicle weight was 842 [kg] with a wide variation from 550 and 1,060 kg. As a result, the FC differs owing to differences in running resistance (attributed mainly to differences in vehicle shape), transmission type, drive system, and turbocharging, which result in the low R^2 value (0.471).

	LP-GV		P-DV		P-GV		P-HV	
n	265		113		1,636		8	
R^2	0.471		0.336		0.700		0.925	
	$b \times 10^{-3}$	c	$b \times 10^{-3}$	c	$b \times 10^{-3}$	c	$b \times 10^{-3}$	c
B (95 CI)	9.09 (7.92–10.3)	0.493 (-0.445 –1.43)	5.10 (3.75 –6.45)	2.95 (0.433 –5.47)	8.45 (8.18 –8.72)	0.446 (0.0792 –0.812)	4.64 (3.32 –5.96)	0.238 (-1.57 –2.04)
t	15.3	1.03	7.49	2.32	61.8	2.39	8.60	0.322

Table 3. Estimates of parameters by Equation 6. n is sample number, B is partial regression coefficient and t is t statistics, respectively.

Of the 113 P-DVs plotted in Figure 7, 25 are 4WD, 92 have AT/CVT transmission, 108 are turbocharged, and 11 have a direct injection engine. The vehicle weight range of 1,705—2,165 [kg] is small compared with that of P-GVs (715—2,380 [kg]). The low R^2 value (0.336) for P-DVs indicate that it is difficult to explain actual FC only with vehicle weight, for the actual FC of a vehicle varies by the combinations of various vehicle specifications.

4.3. Effect of vehicle technologies on actual FC of gasoline-fuelled passenger vehicles

A multiple regression analysis was conducted to evaluate the effect of vehicle technologies on the actual FC of gasoline-fuelled passenger vehicles (P-GVs and P-HVs). A P-GV with a manual transmission and 2WD was set as the baseline. The regression equation can be described as Equation 7:

$$\begin{aligned} FC_{v,actual} &= d_0 + d_1 w + d_2 D_{HV} + d_d D_{AT/CVT} \\ &\quad + d_4 D_{TC} + d_5 D_{4WD} + d_6 D_{DI} + d_7 D_{VVT} \end{aligned} \qquad (7)$$

where w is vehicle weight [kg] and D_{HV}, $D_{AT/CVT}$, D_{TC}, D_{4WD}, D_{DI}, and D_{VVT} are the dummy variables for P-HV, transmission (AT/CVT), turbocharging (TC), 4WD, direct injection (DI), and variable valve timing (VVT), respectively. The parameter estimates are summarized in Table 4.

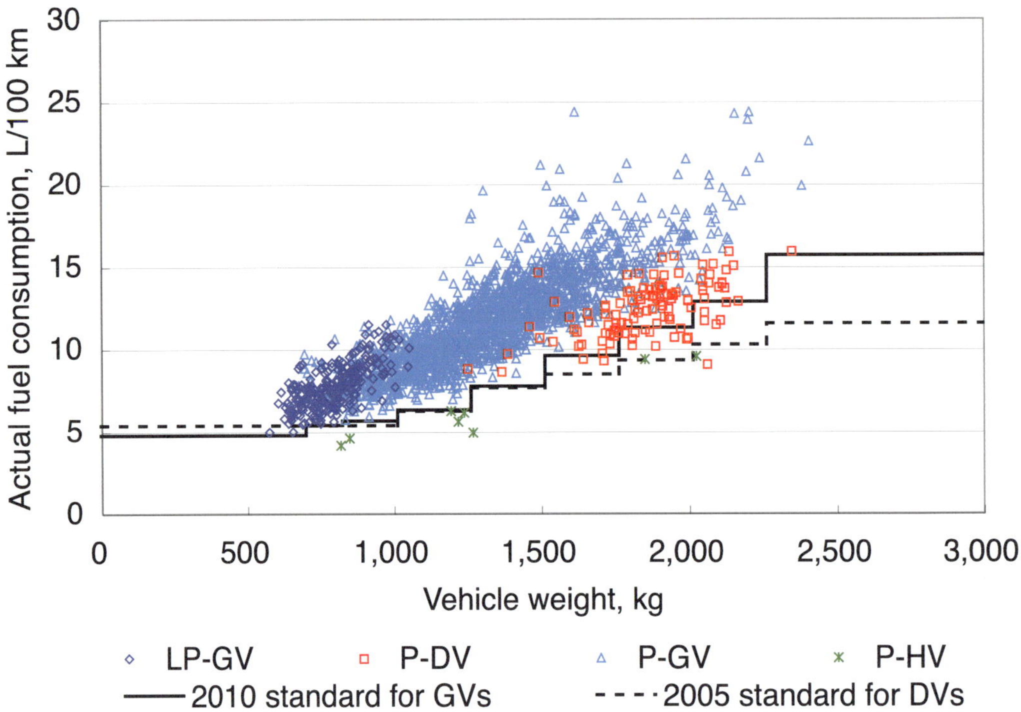

Figure 7. Vehicle weight and actual FC

Using the estimates shown in Table 4 and Equation 7, it is confirmed that the use of direct injection and variable valve timing led to a decrease in actual FC, whereas the use of an automatic transmission and turbocharging resulted in an increase in actual FC. Although the partial regression coefficients of HV and 4WD are negative, adding hybrid technology and 4WD to a baseline P-GV increased vehicle weight. Hence, to evaluate the effect of hybridisation and 4WD, the balance between vehicle weight increase and the coefficients of the dummy variables given in Table 4 should be considered.

Among the 8 HV models included in the actual FC database, 3 models also had equivalent GVs within the same vehicle name, 3 had engines that were variants of the GV models, and 2 were dedicated HV models. Therefore, counterpart GV models could be defined for 6 of the 8 HV models. Although the vehicle weight of HVs depends upon various vehicle specifications, the weight increase of these 6 HVs from their counterpart GVs ranged from 40 to 195 [kg]. Equation 7 and Table 4 were then used to estimate a 0.336—1.64 [L/100km] increase in actual FC from hybridisation. Because the actual FC improvement effect evaluated from the partial regression coefficient of HV prevailed in this estimate, however, it is estimated that hybridisation contributed to an actual FC improvement (-4.44 to -3.14 [L/100km]) from the baseline P-GV.

Of the 1,615 samples analysed in this section, 370 had the same vehicle name and model year for both 2WD and 4WD models (other specifications, such as transmission type, turbocharging, and direct injection, were the same). The use of 4WD increased weight by

83.1 [kg] on average (standard deviation = 38.8 [kg]). The partial regression coefficients d_1 and d_5 shown in Table 4 indicate that a weight increase of 83.1 kg would result in an actual FC increase of 0.329 [L/100km].

n	1,615							
R^2	0.799							
	d_0	$d_1 \times 10^{-3}$	d_2	d_3	d_4	d_5	d_6	d_7
B (95 CI)	0.487 (0.206 – 0.768)	8.39 (8.16 – 8.63)	-4.77 (-5.71 – -3.83)	0.532 (0.379 – 0.685)	0.952 (0.777 – 1.13)	-0.368 (-0.520 – -0.217)	-1.01 (-1.41 – -0.699)	-1.02 (-1.18 – -0.869)
t	3.40	70.9	-9.98	6.83	10.6	-4.77	-5.81	-13.0

Table 4. Estimates of parameters by Equation 7. n is sample number, B is partial regression coefficient and t is t statistics, respectively.

5. Annual differences in mean actual FC of gasoline-fuelled passenger vehicles

Annual (fiscal year, FY) changes in vehicle weight and actual FC for gasoline-fuelled passenger vehicles from FY 2001 to 2004 are analysed. Table 5 presents the descriptive statistics of vehicle weight for gasoline-fuelled passenger vehicles (P-GVs and P-HVs) that were used to conduct a one-factor analysis of variance. No significant difference was observed for mean vehicle weight of P-HVs ($F = 0.252$, $p = 0.859$), but a significant difference was found for P-GVs ($F = 2.71$, $p = 0.044$). Therefore, a post-hoc multiple comparison by Sheffé's test on vehicle weight of P-GVs was conducted, but no significant differences were observed.

Similarly, the mean differences of actual FC are tested. As shown in Section 4.2, actual FC is presented as proportional to vehicle weight; therefore, an analysis of covariance was carried out to adjust for the effect of vehicle weight in actual FC. Mean actual FC of P-GVs decreased significantly from FY2001 until FY2004 ($F = 19.7$, $p = 0.000$). Post-hoc multiple comparisons with the Sidak adjustment showed that the mean actual FC values adjusted for vehicle weight were significantly different, except between FY2003 and FY2004 (Table 6). No significant differences were observed for P-HVs ($F = 0.299$, $p = 0.826$).

The results indicate that the actual FC of P-GVs included in the actual FC database steadily improved, most likely as a result of an increase in the number of vehicles equipped with FC-improving technologies and not because of weight reductions. The lack of significant changes for P-HVs can be attributed to the fact that only small numbers of new-type P-HVs had entered the Japanese passenger vehicle fleet at the time of the study and also to a lack of drastic improvements in the P-HVs produced during this period.

FY	P-GV			P-HV		
	n	μ	σ	n	μ	σ
2001	970	1,330.92	268.04	5	1,196.00	415.73
2002	1,073	1,342.16	275.34	4	1,290.00	414.17
2003	1,091	1,353.79	272.29	7	1,378.57	410.15
2004	1,089	1,362.95	269.70	7	1,378.57	410.15
All	4,223	1,347.94	271.70	23	1,323.48	390.39

Table 5. Descriptive statistics of vehicle weight [kg]. n is the number of vehicle types, μ is the population mean, and σ is standard deviation.

[L/100km]	FY2002	FY2003	FY2004
FY2001	0.190*	0.403*	0.441*
FY2002		0.213*	0.251*
FY2003			0.038

Table 6. Results of post-hoc multiple comparisons using the Sidak adjustment for the mean actual FC of P-GVs. The value in each cell shows the differential in the population mean μ_r in rowwise group r and μ_c in columnwise group c. For example, $\mu_{FY2001} - \mu_{FY2002} = 0.190$. Asterisk denotes significance at 5% level.

6. Validity of actual FC obtained from the actual FC database

To check the validity of the actual FC values calculated from the database, two cases from the database were compared with a third that was calculated from published statistics for gasoline-fuelled passenger vehicles (P-GVs and P-HVs):

Case A: The actual FC of gasoline-fuelled passenger vehicles was estimated for each FY directly from the database.

Case B: The actual FC of gasoline-fuelled passenger vehicles was estimated from the results of the regression analysis between vehicle weight and actual FC (Table 7, Equation 6) and the estimated number of vehicles owned, by vehicle weight (by 10 kg increments), for each FY.

Case C: The actual FC of gasoline-fuelled passenger vehicles was estimated from national statistics (MLIT, 2003–2005).

For Case B, the number of vehicles owned by vehicle weight was estimated from the vehicle specification database and various published statistics (AIRIA1, 2003–2005; AIRIA2, 2003–2005; AIRIA3, 2003–2005). Figure 8 shows the ownership rate (OR) relative to the total number of vehicles owned, by Japanese inertia weight class, for passenger vehicles from FY2002 (March 2003) to FY2004 (March 2005). The sampling rate (SR) — the number of vehicles actually included in the estimates of Case A as a ratio of the total number of vehicles owned – is also shown in the figure. The vehicle weight distribution of the database (SR) does not reflect the real-world distribution (OR); OR has a normal distribution, whereas SR is higher for both light (< 702 [kg]) and heavy (1,766+ [kg]) vehicles. Therefore, actual FC values compiled directly from the database in Case A might have been biased as a result of the vehicle weight distribution.

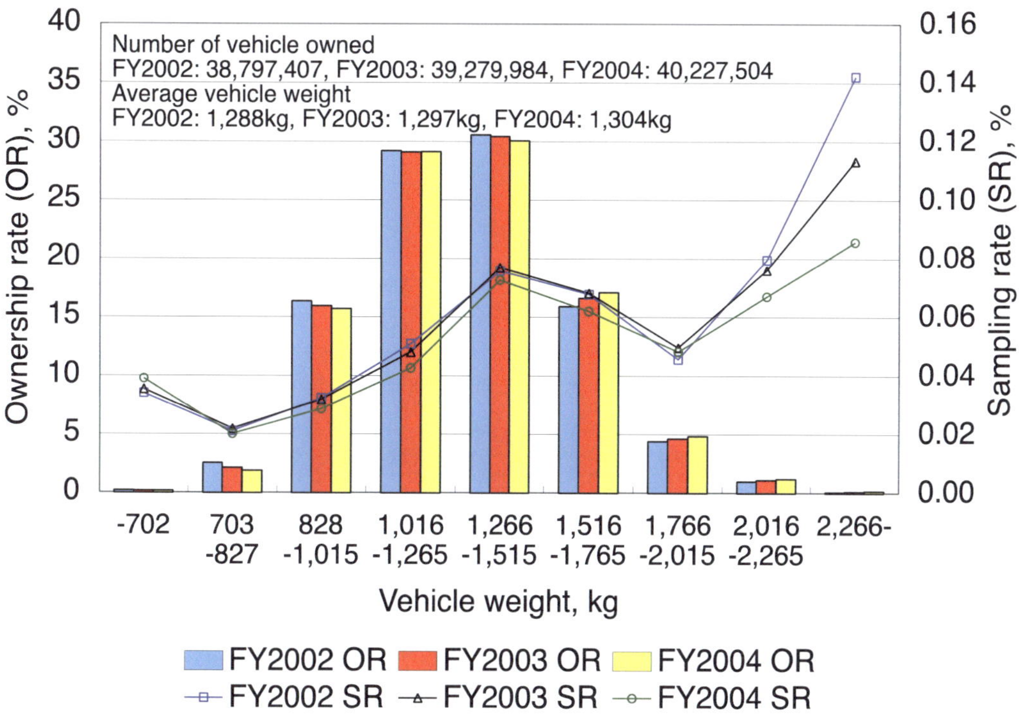

Figure 8. Ownership rates of gasoline-fuelled passenger vehicles and sampling rates of vehicles included in the database.

As described in Section 4, the mean actual FC adjusted by vehicle weight improved each year in the study period. Therefore, Case B was designed to reflect improvements in actual FC adjusted for the vehicle weight bias that might have been included in the database (Case A). Because no significant improvement in actual FC was observed for P-HVs from FY2002 to FY2004, the results of the regression analysis shown in Table 3 were used for P-HVs; the results from Table 7 were used for P-GVs.

Table 8 shows the estimates of actual FC of gasoline-fuelled passenger vehicles for the three cases from FY2002 to FY2004. The actual FC steadily improved from FY2002 to FY2004 in Cases A and C, but the actual FC did not improve from FY2003 and FY2004 in Case B, similar to the results in the same time period shown in Table 6. Although there are small differences in each FY, the estimates of actual FC of gasoline-fuelled passenger vehicles in Cases A and B were within 4% of the Case C estimates in all instances. Therefore, the actual FC values derived from the database appear to be compatible with the estimates from published statistics.

	FY2002		FY2003		FY2004	
n	1,073		1,091		1,089	
R^2	0.729		0.704		0.684	
	$b \times 10^{-3}$	c	$b \times 10^{-3}$	c	$b \times 10^{-3}$	c
B (95 CI)	8.57 (8.25 – 8.88)	0.338 (-0.0910 – 0.767)	8.38 (8.06 – 8.70)	0.377 (0.0692 – 0.822)	8.33 (7.99 – 8.67)	0.408 (0.0599 – 0.876)
t	53.7	1.55	50.9	1.66	48.5	1.71

Table 7. Estimates of parameters by Equation 4 for P-GVs for FY2002–2004. n is sample number, B is partial regression coefficient and t is t statistics, respectively.

Case or comparison	FY2002	FY2003	FY2004
Case A	11.81	11.62	11.59
Case B (95CI)	11.36 (11.24 – 11.48)	11.23 (11.10 – 11.36)	11.24 (11.11 – 11.37)
Case C	11.66	11.49	11.13
Case A / Case C	1.01	1.01	1.04
Case B / Case C (95 CI)	0.974 (0.964 – 0.984)	0.977 (0.966 – 0.988)	1.01 (0.998 – 1.02)

Table 8. Comparison of actual FC [L/100km] of gasoline-fuelled passenger vehicles for Cases A–C.

7. Conclusion

In order to quantify the relationship between vehicle specifications and actual FC with statistical reliability, an actual FC database was developed by using vehicle specification data and voluntarily reported data collected from an internet-connected mobile phone system throughout Japan. The database was used to conduct statistical analyses to evaluate the effects of various vehicle specifications on the FC/FE of passenger vehicles. The actual FC adjusted by vehicle weight was shown to have significantly improved from FY2001 to FY 2004. Moreover, estimates of the actual FC of gasoline-fuelled passenger vehicles obtained from the database were consistent with estimates calculated from national statistics.

With the revision of the Energy-Saving Law in July 2007, Japan changed from using the 10-15 mode to the JC08 mode (UNEP, 2012); the new 2015 FE standards for passenger vehicles are based on the Top Runners Approach provided in the JC08 mode. Japanese vehicle makers have already started to sell new passenger vehicles that have achieved the 2015 FE standard, so the effects of equipping vehicles with various types of new and more fuel efficient technologies may influence the actual FC of these vehicles as well. The author's group plans to extend the data collection period presented in this paper and to update the actual FC database to reflect state-of-the-art vehicle technologies in the real world.

Finally, the World Forum for Harmonization of Vehicle Regulations, which is a working party (WP.29) of the United Nations Economic Commission for Europe, has decided to set up an informal group under its Working Party on Pollution and Energy to develop a worldwide harmonized light duty test cycle (the Worldwide Harmonized Light Duty Vehicle Test Procedures, WLTP) by 2013. This cycle will represent typical driving conditions around the world (UNECE, 2012). Because the actual FC/FE of vehicles might show different trends if the WLTP is adopted and applied to meet new FC/FE standards, the movement towards the endorsement of the WLTP could influence future studies as well.

Author details

Yuki Kudoh
Research Institute of Science for Safety and Sustainability,
National Institute of Advanced Industrial Science and Technology, Japan

8. References

An F., Earley, R. & Green-Weiskel, L. (May 2011). Global Overview on Fuel Efficiency and Motor Vehicle Emission Standards: Policy Options and Perspectives for International Cooperation, United Nations Commission of Sustainable Development, Background Document CSD19/2011/BP3. Retrieved from <http://www.un.org/esa/dsd/resources/res_pdfs/csd-19/Background-paper3-transport.pdf>

Automobile Inspection & Registration Information Association (AIRIA1) (2003–2005). *Number of Vehicles Owned by Vehicle Weight as of March Each Year,* (in Japanese)

Automobile Inspection & Registration Information Association (AIRIA2) (2003–2005). *Number of Vehicles Owned by Vehicle Weight by Engine Displacement Class as of March Each Year,* (in Japanese)

Automobile Inspection & Registration Information Association (AIRIA3) (2003–2005). *Number of Vehicles Owned by Vehicle Weight by Vehicle Type as of March Each Year,* (in Japanese)

Duoba, M., Lohse-Bush, H. & Bohn, T. (2005). Investigating Vehicle Fuel Economy Robustness of Conventional and Hybrid Electric Vehicles, *Proceedings on the 21st Worldwide Battery, Hybrid and Fuel Cell Electric Vehicle Symposium & Exhibition,* Monaco, April 2005

Farrington, R. & Rugh, J. (October 2000). *Impact of Vehicle Air-Conditioning on Fuel Economy, Tailpipe Emissions, and Electric Vehicle Range: Reprint,* National Renewable Energy Laboratory, Retrieved from <http://www.nrel.gov/docs/fy00osti/28960.pdf>

Huo H, Yao Z, He K, Yu X (2011) Fuel Consumption Rates of Passenger Cars in China: Labels Versus Real-world. *Energy Policy,* Vol. 39, Issue 11, (November 2010), pp. 7130–7135, ISSN 0301-4215

Kudoh, Y., Ishitani, H., Matsuhashi, R., Yoshida, Y., Morita, K., Katsuki, S. & Kobayashi, O. (2001). Environmental Evaluation of Introducing Electric Vehicles Using a Dynamic Traffic Flow Model, *Applied Energy,* Vol. 69, Issue 2, (June 2001), pp. 145-159, ISSN 0306-2619

Kudoh, Y., Kondo, Y., Matsuhashi, K., Kobayashi, S. & Moriguchi, Y. (2004). Current status of actual fuel-consumptions of petrol-fuelled passenger vehicles in Japan, *Applied Energy*, Vol. 79, Issue 3, (November 2004), pp. 291-308, ISSN 0306-2619

Kudoh, Y., Matsuhashi, K., Kondo, Y., Kobayashi, S. Moriguchi, Y. & Yagita, H. (2007). Statistical Analysis of Fuel Consumption of Hybrid Electric Vehicles in Japan, *The World Electric Vehicle Association Journal*, Vol. 1, pp. 142-147, ISSN 2032-6653

Kudoh, Y., Matsuhashi, K., Kondo, Y., Kobayashi, S. Moriguchi, Y. & Yagita, H. (2008). Statistical Analysis on the Transition of Actual Fuel Consumption by Improvement of Japanese 10•15 Mode Fuel Consumption, *Journal of the Japan Institute of Energy*, Vol. 87, No. 11, (November 2008), pp. 930-937, ISSN 0916-8753, (in Japanese)

Ministry of Land, Transport and Infrastructures (MLIT) (2003–2005). *Annual Statistics of Automobile Transport*, (in Japanese)

Nishio, Y., Kaneko, A., Murata, Y., Daisho, Y., Sakai, K. & Suzuki, H. (2008). Consideration of Evaluation for Fuel Consumption under using Air Conditioner, *Transactions of Society of Automotive Engineers of Japan*, Society of Automotive Engineers of Japan, Vol. 39, No. 6, (November 2008), pp. 6_229-6_234, ISSN 0287-8321, (in Japanese)

Sagawa, N. & Sakaguchi, T. (2000). Possibility of introducing fuel-efficient vehicles and fuel consumption trends of passenger vehicles, *Proceedings on the 16th Conference on Energy System, Economy, and the Environment*, Japan Society of Energy and Resources, Tokyo, January 2000, (in Japanese).

Schipper, L. & Tax, W. (1994). New car test and actual fuel economy: yet another gap? *Transport Policy*, Vol. 1, Issue 4, (October 1994), pp. 257-265, ISSN 0967-070X

Schipper, L. (2011). Automobile use, fuel economy and CO2 emissions in industrialized countries: Encouraging trends through 2008? *Transport Policy*, Vol. 18, Issue 2, (March 2011), pp. 358-372, ISSN 0967-070X

Tokyo Metropolitan Government Bureau of Environment (TMGBE) (1996). *Investment Report of Traffic Volume and Exhaust Gases from Vehicles (Outline)*, (in Japanese).

United Nations Economic Commission for Europe (UNECE) (n.d. 2012). Working Party on Pollution and Energy (GPRE), In: *UNECE*, 28.03.2012, Available from: < http://www.unece.org/trans/main/wp29/meeting_docs_grpe.html>

United Nations Environment Programme (UNEP) (2012). Japanese Test Cycles, In: *Cleaner, More Efficient Vehicles*, 28.03.2012, Available from: <http://www.unep.org/transport/gfei/autotool/approaches/information/test_cycles.asp# Japanese>

United States Environmental Protection Agency (USEPA) (2010). *Light-Duty Automotive Technology, Carbon Dioxide Emissions, and Fuel Economy Trends: 1975 through 2010*, USEPA, Retrieved from <http://www.epa.gov/otaq/cert/mpg/fetrends/420r10023.pdf>

Wang, H., Fu, L, Zhou, Y. & Li, H. (2008). Modelling of the fuel consumption for passenger cars regarding driving characteristics, *Transportation Research Part D: Transport and Environment*, Vol. 13, Issue 7, (October 2008), pp.479-482, ISSN 1361-9209